the FUTURE isWILD™

DOUGAL DIXON · JOHN ADAMS

FIREFLY BOOKS

A FIREFLY BOOK

Published by Firefly Books Ltd., 2003

First Printing

National Library of Canada Cataloguing in Publication Data

Dixon, Dougal
        The future is wild / Dougal Dixon, John Adams.

Includes index.
ISBN 1-55297-724-2 (bound).—ISBN 1-55297-723-4 (pbk.)

    1. Evolution (Biology)—Forecasting. 2. Evolution
(Biology)—History. I. Adams, John II. Title.

QH371.3.F8D59 2003   576.8   C2002-903545-7

Publisher Cataloging-in-Publication Data (U.S.)

Dixon, Dougal.
   The future is wild / Dougal Dixon, and John Adams. —1st ed.
[160] p. : col. ill. : col. photos. ;   cm.
Includes index.
Note: Companion book to the Discovery Channel series.
Summary: How life on Earth may evolve over the next 200 million
years. Written with a team of international scientists, based on
biological and evolutionary principles.
ISBN 1-55297-724-2
ISBN 1-55297-723-4   (pbk.)
1. Evolution (Biology). 2. Life. I. Adams, John. II. Title.
576.7 21   QH367.1.D59 2003

Published in Canada in 2003 by
Firefly Books Ltd.
3680 Victoria Park Avenue
Toronto, Ontario M2H 3K1

Published in the United States in 2003 by
Firefly Books (U.S.) Inc.
P.O. Box 1338, Ellicott Station
Buffalo, New York 14205

Reproduction by Colourscan, Singapore
Printed in Spain

**This book and the associated television series were
conceived by** The Future is Wild Ltd, Solomon's Court,
Bournes Green, Stroud GL6 8LY

This book was created by act-two, 346 Old Street,
London EC1V 9RB. www.act-two.com

Text consultants
Professor R McNeill Alexander, University of Leeds, UK
Professor Bruce Tiffney, University of California, USA

Researchers John Capener; Belinda Biggam

Managing editor Claire Pye; **Senior designer** Abigail Hicks;
Editorial support Mark Blacklock, Brian Muir, Paul Virr;
Picture research Ellen Root; **Illustrations** Peter Bull Art Studio,
Mel Pickering; **Image retouching** Itchy Animation; **Digital
artwork** 422; **Editorial director** Jane Wilsher; **Art director**
Belinda Webster; **Production** Adam Wilde; **Index** Ann Barrett

The television series was created by
The Future is Wild Ltd

Series producer Paul Reddish; **Series director** Steve Nicholls;
Series writer Victoria Coules; **Animation director** Peter Bailey;
Producers Jeremy Cadle, Clare Dornan; **Production manager**
Wolfgang Knöpfler; **Picture researcher** Lawrence Breen;
Editors Liz Thoyts, Martin Elsbury; **Composers** Nick Hooper,
Paul Pritchard; **Production coordinator** Kensa Duncan

The Future is Wild would not have been possible
without the support of the following consultants:

Professor R McNeill Alexander
Professor Emeritus of Zoology, University of Leeds, UK

Dr Letitia Aviles
Associate Professor, Department of Ecology and
Evolutionary Biology, University of Arizona, USA

Dr Phillip Currie
Head of Dinosaur Research Program and Curator of Dinosaurs and Birds,
Royal Tyrell Museum of Paleontology, Canada

Professor Richard Fortey
Department of Paleontology,
The Natural History Museum, UK

Professor William Gilly
Professor of Cell and Developmental Biology
and Marine Biology, Stanford University, USA

Professor Stephen Harris
Mammal Research Unit, University of Bristol, UK

Mike Linley
Herpetologist, Hairy Frog Productions, UK

Dr Roy Livermore
British Antarctic Survey, UK

Professor Karl Niklas
Liberty Hyde Bailey Professor of Plant Biology,
Cornell University, USA

Professor Stephen Palumbi
Professor of Biology, Stanford University, USA

Professor Jeremy Rayner
Alexander Professor of Zoology, University of Leeds, UK

Professor Bruce Tiffney
Professor of Geological Sciences,
University of California, USA

Professor Paul Valdes
Department of Meteorology, Reading University, UK

CONTENTS

Evolving Earth

To imagine the future, we must first look to the past. By tracing the history of life on Earth, we can begin to see the recurring patterns of evolution that will help us predict what the future may hold.

5 Million Years

Earth is now at the peak of an Ice age that began well before human times. Northern Europe and North America are covered by ice sheets. The world is a cold dry place where only the hardiest, most adaptable species are able to survive.

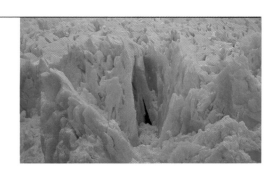

100 Million Years

Earth has enjoyed a long period of stable conditions since the last Ice age, and life has bounced back. The icecaps have melted, sea levels have risen and the world is warm and humid. It is a global hothouse, brimming with life.

200 Million Years

The planet has changed. A single, huge supercontinent shares Earth with a vast, warm ocean. It is 100 million years since the mass extinction that destroyed 95 percent of species on Earth. But evolution is probably at its most inventive after a mass extinction.

FOREWORD

Take a walk in a forest anywhere in the world today and you'll see the same basic types of animals and vegetation – birds, mammals and flowering plants. But a forest two hundred million years ago was a very different place. There were no birds. Mammals and flowering plants had only just begun to evolve. Who would have thought that, in the last 200 million years – a mere fraction of the time that life has graced our planet, so many completely new organisms would have evolved and thrived?

So what does evolution have in store for life on Earth over the next 200 million years? What creatures will roam the land or swim in the oceans? The Future is Wild draws on an international team of experts who bring to life a world of amazing organisms, setting them loose in our imaginations. But the rules are strict. The organisms you'll encounter in this book are based on fundamental biological and evolutionary principles. They could, and may yet, exist.

The time is right to consider the future. Using state-of-the-art computer animations, The Future is Wild has been able to transform imagination into images, creating a living world of strange creatures and extraordinary habitats. The future you are about to see is very wild indeed, and the creatures that populate the pages of this book are just a few of the possibilities...

left

The eight-ton megasquid, evolved from a marine squid, is just one of the creatures to roam the amazing world of The Future is Wild.

Stephen Palumbi is Professor of Biology at Stanford University.

IMAGINING THE FUTURE

Professor R. McNeill Alexander is Professor Emeritus of Zoology at the University of Leeds, UK. He is a specialist in biomechanics, the study of animal movement. He has provided invaluable help and advice on many aspects of the animals and habitats featured in The Future is Wild television series. Here, he talks about the scientific processes and methods which underpin this vision of the future.

The Future is Wild tells of eight-ton squids roaming the land in a world where all the continents have merged into a single, huge landmass. It tells of snails that hop like kangaroos, fish that fly like butterflies through forests, and birds with four wings. This future world may seem incredible, but it is firmly grounded in science.

Imagining the planet Earth five, 100 and 200 million years in the future is no easy task. In order to bring the habitats and creatures of tomorrow to life, the producers of The Future is Wild have worked closely with an international team of scientific advisers to ensure that everything that is presented is possible.

We began by imagining how Earth's continents might be distributed in the future. For this, we called on an earth scientist. By studying rock magnetism, earth scientists have discovered how, over the past several hundred million years, the continents have slowly moved, regrouped and crushed together to form mountain ranges. There have been no sudden changes of direction in these movements, and our consultant expects them to continue more or less as our future world maps show (see page 13). The position of landmasses and mountains also determine the climates of the future. By studying the world maps, a climatologist was able to deduce the climates of our future habitats.

We have had advice from a great many biologists. Some are acknowledged experts on particular groups of organisms, while others are known for their breadth of knowledge in fields such as ecology, biomechanics and physiology. But even with this fund of knowledge, we have to accept that our predictions, though rooted in science, will necessarily contain some conjecture. In our rich, diverse world, there are simply too many species interacting with each other and with their environments in subtle, complicated ways. The theory of chaos tells us that it is impossible to make reliable long-term predictions for highly complex systems.

Despite the difficulties involved, we have done our best to ensure that the plants and animals of our future worlds are viable, and could evolve from existing species in the time available. Our team of biologists has suggested many remarkable possibilities, such as the megasquid, a giant terrestrial squid living in the Northern Forest 200 million years from now. This animal is the result of detailed advice and calculations from an expert in squids and a specialist in biomechanics.

Our suggestions are based on certain assumptions. We have assumed that the plants and animals of the future will be made of similar materials to present-day plants and animals. For example, we have assumed that wood and bone will be as strong as present-day wood and bone, and that muscles will exert about the same force as present-day muscles of equal size. We have also assumed that the maximum rates of animal growth or photosynthesis in plants, will be no faster than at present. These assumptions have been applied in numerous calculations, to check, for example, that a large land animal would be strong enough to support its own weight or that a flying animal could carry enough fat to fuel its journeys.

A few simple rules, based on observations of present-day animals, have helped us to work out what the lives of future animals would be like. One general rule states that an animal sixteen times heavier than a close relative will need about eight times more food each day and take about twice as long to grow to maturity. The large animals that we imagine might inhabit the world in the future have all been created with this in mind.

In the course of evolutionary history, several amazing patterns of change have occurred repeatedly. We can expect to see similar patterns in the future, in other groups of animals and plants. For example, birds, bats, insects and pterosaurs have all, separately, evolved the ability to fly. Amphibians, lungfishes, land snails, crabs and insects have evolved the ability to breathe air instead of water. Ostriches and certain salamanders have evolved to become sexually mature while retaining juvenile characteristics – with its fluffy feathers and rudimentary wings, an ostrich resembles an overgrown chicken. Aphids, water fleas and rotifers all reproduce by parthenogenesis, or virgin birth. In creating the plants and animals of tomorrow, we have used our knowledge of the past to help us imagine the future.

Birth of a megasquid

How could a marine squid evolve into the giant, terrestrial megasquid? (See pages 148-149.) According to our experts, it is not difficult to imagine that squids may one day live on land, since all land-living animals are descended from marine organisms. Once this possibility was established, our team of biologists calculated how big the megasquid could grow, and how it would support its weight.

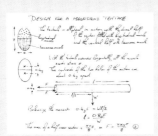
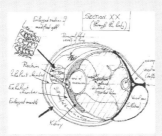

Above are some of the calculations and diagrams which went into the making of a megasquid.

Scientists believe that, 200 million years in the future, the eight-ton megasquid may roam Earth.

EVOLVING
EARTH

DYNAMIC EARTH

According to scientific estimates, Earth has taken more than 4,550 million years to evolve from a mass of dust and gas into the planet we live on today. There is evidence that life has existed on the planet for around 3,800 million years. In all that time, Earth's surface has been in a constant state of flux.

We take it for granted that we are walking on solid ground, yet events frequently occur that give away Earth's dynamic nature. Earthquakes bring devastation to towns and cities; ocean waves nibble away at cliffs; floods cause rivers of rock and soil to slide down mountains. These are geological changes that take place within our lifetimes.

Earth inside out

The secret of Earth's dynamic nature lies in its composition. It is made up of many different layers.

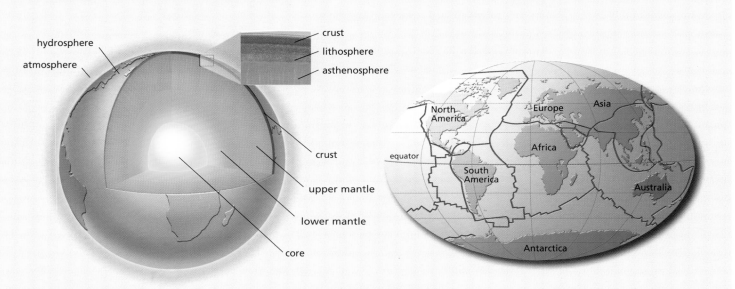

Earth's layers
The outermost layer of Earth is the atmosphere, the air we breathe. Next is the hydrosphere – the world's oceans and rivers. The ground we walk on and the ocean basins are the crust, a layer of lightweight rocks. The upper mantle consists of several other layers of rock. Rocks close to the crust form the lithosphere, Earth's brittle surface. Below, lies the asthenosphere, a layer of molten rocks. The mantle and core are the dense, heavy interior of Earth.

Earth's plates
The continents of Earth sit on top of giant plates of lithosphere, forming a vast jigsaw puzzle on the planet's surface. The plates float on the asthenosphere, the hot, viscous layer of rocks below. Heat energy from deep within Earth generates vast, slow-moving convection currents, causing the asthenosphere to flow very slowly. This movement gradually drives the plates across the surface, carrying the continents with them.

Continental drift

Over millions of years, the continents of Earth drift apart, jostle together and regroup to form new landmasses. By tracking the movement of the continents from the past to the present, scientists have been able to predict how the surface of the planet might look in the future.

225 million years ago
By the Triassic period, most of the continents had joined, forming a single landmass called Pangaea.

80 million years ago
The break up of Pangaea opened up the Atlantic Ocean and caused the Pacific Ocean to shrink. Australia began to move closer to the equator.

Earth today – Human era
Today, there are seven continents and many oceans and seas. The landmasses are widely dispersed. Species evolve separately in different climate zones, leading to great diversity.

5 million years in the future
Earth is in the grip of an Ice age. Much of the northern hemisphere is covered by ice and sea levels have dropped, causing the Mediterranean Sea to dry up.

100 million years in the future
The Ice age has ended and sea levels have risen, changing the coastlines of Earth. Australia has collided with Asia. Antarctica has moved north into a warmer climate zone.

200 million years in the future
The separate continents have become one – a sprawling supercontinent called Pangaea II. This vast landmass shares the planet with a single ocean, known as the Global Ocean.

Other geological changes take place over millions of years, gradually shaping the landscapes of the world. Mountains, volcanoes and valleys are all forged by the movement of Earth's plates as they slowly drift across the globe. The theory of plate motion, and the geological changes it brings about, is known as plate tectonics.

The plates are driven by vast, slow-moving convection currents in the semi-fluid layer of the asthenosphere. These currents bring molten lithosphere to the surface, where it becomes attached to the edge of a plate, and solidifies into new crust. Along the opposite edge, old crust is dragged back down into the asthenosphere and melted. This constant recycling action, like a giant conveyor belt, means that Earth's crust is never more than 200 million years old. New crust is created along deep cracks in the ocean floor, known as mid-ocean ridges, where plates are slowly moving apart. In places, these ridges emerge on to land, splitting continents and forming long rift valleys.

Each plate forms along one edge and moves past other plates at a rate of inches per year. Where the plates eventually come together, along other edges, old crust is dragged down into the asthenosphere and destroyed. This process, known as subduction, creates deep ocean trenches alongside many of Earth's continents. As old crust is melted, new magma is produced, erupting on the surface in the form of volcanoes.

On the surface of the plates lie masses of lighter rock. These rocks form the continental crust, which is too light to be easily dragged down during subduction. Carried into collision by plate movements, the rocks pile up on one another, throwing up great mountain ranges. But even as mountains are formed, the forces of erosion and gravity are acting on them. Over time, even the highest mountains will be worn down to sea level once more. The rocky peaks will be broken into sand and rubble, washing down to the sea to become the basis for new continental material.

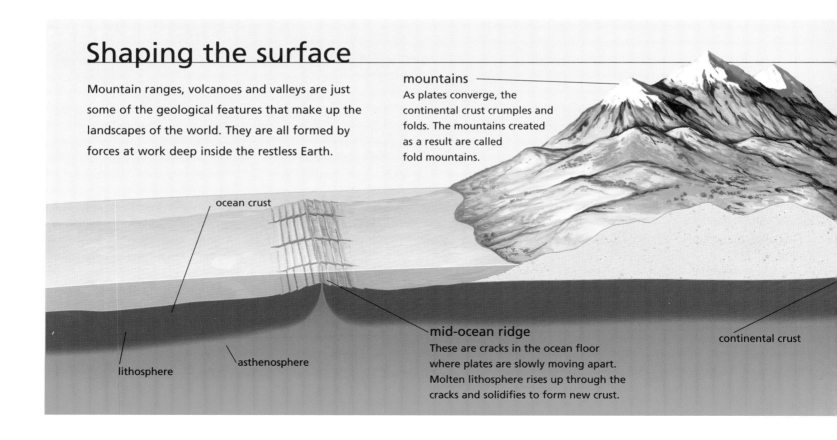

Shaping the surface

Mountain ranges, volcanoes and valleys are just some of the geological features that make up the landscapes of the world. They are all formed by forces at work deep inside the restless Earth.

mountains
As plates converge, the continental crust crumples and folds. The mountains created as a result are called fold mountains.

ocean crust

lithosphere

asthenosphere

mid-ocean ridge
These are cracks in the ocean floor where plates are slowly moving apart. Molten lithosphere rises up through the cracks and solidifies to form new crust.

continental crust

left
Dawn over Mount Makalu and
Mount Chomolonzo in Tibet. The
Himalayas were formed by the
collision of India against Asia.

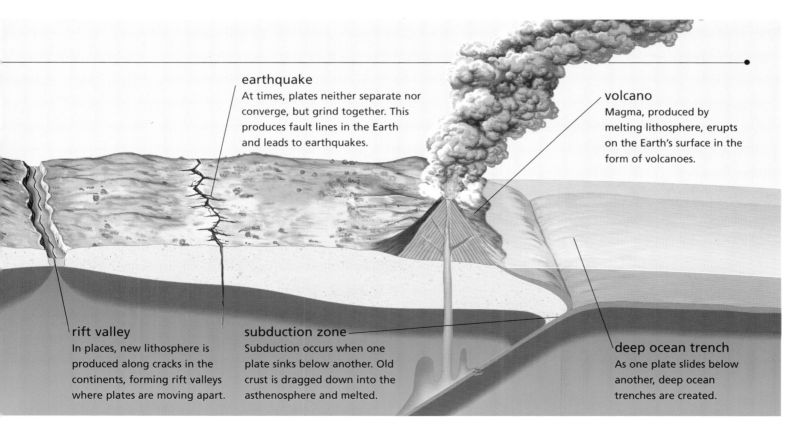

earthquake
At times, plates neither separate nor
converge, but grind together. This
produces fault lines in the Earth
and leads to earthquakes.

volcano
Magma, produced by
melting lithosphere, erupts
on the Earth's surface in the
form of volcanoes.

rift valley
In places, new lithosphere is
produced along cracks in the
continents, forming rift valleys
where plates are moving apart.

subduction zone
Subduction occurs when one
plate sinks below another. Old
crust is dragged down into the
asthenosphere and melted.

deep ocean trench
As one plate slides below
another, deep ocean
trenches are created.

CYCLES OF LIFE

It is easy to think that the study of evolution started when Charles Darwin published The Origin of Species in 1859. In fact, humans have been documenting observations about the living world since cave-dwellers began painting on cave walls. From traditional folk songs to the exhaustive classification of species in the Victorian era, an enormous understanding of life on Earth has been developed by the human race.

Darwin and his contemporaries revolutionized the way we view nature and our place in it. They suggested that different species of animals, despite differences in their appearance and behavior, might share a common ancestor. Suddenly, the history of life was revealed as a sprawling tree, grown from a single seed and slowly spreading its vast canopy of branches through time.

Darwin and his colleagues set out the theory of natural selection, suggesting that organisms which can survive and reproduce in a given environment may pass on their attributes to their offspring, thereby ensuring that these attributes are retained within the species. This theory has been refined a great deal since Darwin's time. Scientists can now study evolution by examining the molecules that make up the cells of living things. Modern science sees natural selection as just one chapter in a complex story.

At the center of evolutionary theory is the gene pool. A gene pool is all the genetic attributes of a species, spread throughout one or many populations. Genes become more or less frequent within a gene pool according to whether their effects are beneficial, neutral or adverse. A gene which causes disease is less likely to be passed on because the organism carrying it is more likely to die before reproducing. Conversely, a gene which increases an

The survival of this warthog depends on a combination of genes which enables it to run at high speeds. It may live to breed and therefore pass on its favorable genes – a process known as natural selection.

animal's speed, and therefore its ability to escape from predators, may help an individual to survive and breed. This gene will be passed on and will survive in the gene pool. Because an organism inherits half of its genetic code from each parent, genetic associations are constantly varied by sexual recombinations.

Evolution occurs when there is a change in a population's gene pool. One such change may be the addition of genes by new individuals as they move into a population from another area. Genes may also be changed by mutation, where a random fault in the genetic code creates a new characteristic. If this characteristic is beneficial, or neutral, it may be passed on to the next generation. The process of adding and subtracting from the gene pool results in almost imperceptible changes from one generation to the next. However, by studying these processes over long time periods in the fossil record, we can see how new species have evolved and then branched or radiated out into different groups.

The importance of basic solutions to survival within a particular environment may result in recurrent patterns of evolution. For example, it may be that a certain body shape is most effective for animals with certain lifestyles. A similar body shape therefore evolves again and again in different animal groups living in a similar environment. This occurrence, known as convergent evolution, can be seen in the way birds and bats have separately evolved the ability to fly. In another process, called exaptation, a feature that has already evolved for one function is adapted for another. An example of exaptation is a penguin's wing, which was once used for flying and is now used for swimming.

Changes in environment are also key to evolution. Should a new food source develop because of a change in ocean currents, new swimming animals will evolve to exploit it. If a new island appears, with no ground-dwelling predators, then certain birds may discard flight to feed on the plants and insects that have colonized the land. The newly-formed island, and the food sources on it, represent an evolutionary niche, ready to be filled by new species. In other words, a niche is the way an animal makes a living from all of the resources in its habitat. Whenever a niche is opened, it is quickly filled. Nature does not leave gaps.

Earth today, in the time period this book will refer to as the Human era, is home to a greater diversity of life than at any other time in its history. This diversity can, in part, be explained by the wide range of habitats on the planet. Desert, tundra, forest, grassland and aquatic habitats all exist in infinite variety across the continents and temperature zones of Earth. And, within each habitat, different species of animal and plant life exploit all the available niches.

Darwin's finches

A favorite example of natural selection is Darwin's finches. During his famous expedition on *HMS Beagle*, he collected the fourteen species of finch on the Galápagos Islands. Each species seemed to have a beak ideally suited to eating its preferred food. After years of detective work, Darwin and his colleagues deduced that all fourteen species must have shared a common ancestor, blown to the islands from South America. This small ancestral finch population must have had a rich gene pool, giving rise to all the possible types of beak, and enabling the finches to exploit the islands' resources. This process is known as adaptive radiation.

The cactus finch has a long, probing beak, ideally suited to plucking insects from a spiny cactus.

No animal exists in isolation in any habitat. Each animal relies on a food source, and that food source derives ultimately from the sun. The sun's energy powers the photosynthesis of plants, allowing them to combine carbon dioxide and water from their surroundings into their own food supplies. These food supplies are eaten by plant-eating animals. Plant-eating animals are eaten by meat-eating animals. All animals and plants die, and their bodies are broken down by bacteria and fungi to produce the carbon dioxide and moisture needed for photosynthesis. This is a cycle.

The sun is the primary source of energy for life, but not the only one. Even in the most remote reaches of Earth, in the ecosystems of hot vents which lie at great depths in the world's oceans, there is life. Here, the original energy input is not from the sun but from the chemistry of Earth's processes.

The whole gamut of life on Earth fits into these cycles: one form of life eats another and in turn forms a food source for a third organism. This structure was once thought of as a food chain. In fact, it is a food web, with numerous interlocking strands. Carnivores often eat other carnivores, plants sometimes eat insects, and so on.

The remora, or sucker-fish, lives in harmony with its shark host. The tiny fish eats scraps of meat left over by the shark as it devours its prey.

Competition between species for resources such as space and food is often a driving force for evolution and adaptation. Termites are one example of species which have evolved a high degree of social organization to ensure that, by working together, they will successfully defend and care for their queen and also the entire colony.

Predation and parasitism, in which one species exploits another, are common patterns in this competitive world. However, there is a great deal of co-operation as well. Symbiosis occurs when two species exist side by side, each benefiting from the presence of the other. One such symbiotic relationship exists between algae and fungi in lichens: the fungi provide a protective structure, while the algae synthesize carbohydrates for food. There is also commensalism, where one species gains an advantage by living off another, but the host species is neither helped nor harmed by the association. The remora, or sucker-fish, attaches itself to the body of a shark. It lives off scraps of food from the shark's prey, while remaining unharmed and even protected by the bigger fish.

Plants need sunlight to survive. Through photosynthesis, plants harness energy in the form of light and convert it to chemical energy for food.

All species ultimately become extinct. Most do so gradually over time, slowly dying out as a result of environmental change and competition. This natural

left
What does the future hold for the human race, with its sprawling cities and vast, polluting industries?

pattern is punctuated by periods where the extinction rate accelerates. Entire groups of organisms are suddenly wiped out in what is known as a mass extinction. This has occurred five times in Earth's past, most recently with the extinction of the terrestrial dinosaurs. On the face of it, such an event is a catastrophe. Millions of years of gradual diversification devastated by a crude blow, randomly eliminating key groups. But for those lucky enough to survive, it may be a golden opportunity. In an established ecosystem, it is almost impossible for a newcomer to gain a foothold. Life can be tough for newly-evolved organisms. However, in the aftermath of a mass extinction, those species waiting in the wings at last have their chance to flourish.

Despite successive mass extinctions, the overall diversity of life on Earth has increased greatly over time. This is perhaps just as well. Humans are currently contributing to a sixth mass extinction. Unlike previous extinctions, which are believed to have been caused by climate change, volcanic

activity or meteors striking Earth, this human-influenced mass extinction is eating away at the habitats of countless populations of plants and animals. It is difficult to predict how the planet will cope with the impact of humans, with our immense consumption of natural resources. Will we, like other species, become extinct? Perhaps humans can survive extinction by colonizing other planets, or by adapting to breathe the air we ourselves have polluted. It may be that we are Earth's ultimate generalists, able to adapt to the most extreme changes in our environment. One thing is certain: to survive, we will need to learn much more about the way our planet works.

But what if our domineering presence was removed, and the rest of the natural world was left to its own devices? What evolutionary success stories are waiting to be told? The Future is Wild charts the evolution of a world without people, a world in which the forces of genetic variation and natural selection, not industry, decide the fate of the planet and the creatures in it.

Life lines

Although Earth was formed around 4,550 million years ago, the earliest fossils of multicellular organisms come from the Vendian period, 4,000 million years later. It has been estimated that Earth still has some 15,000 million years to go before its end.

Geological timeline

The study of rocks and fossils has enabled scientists to piece together the major events that tell the story of life on Earth.

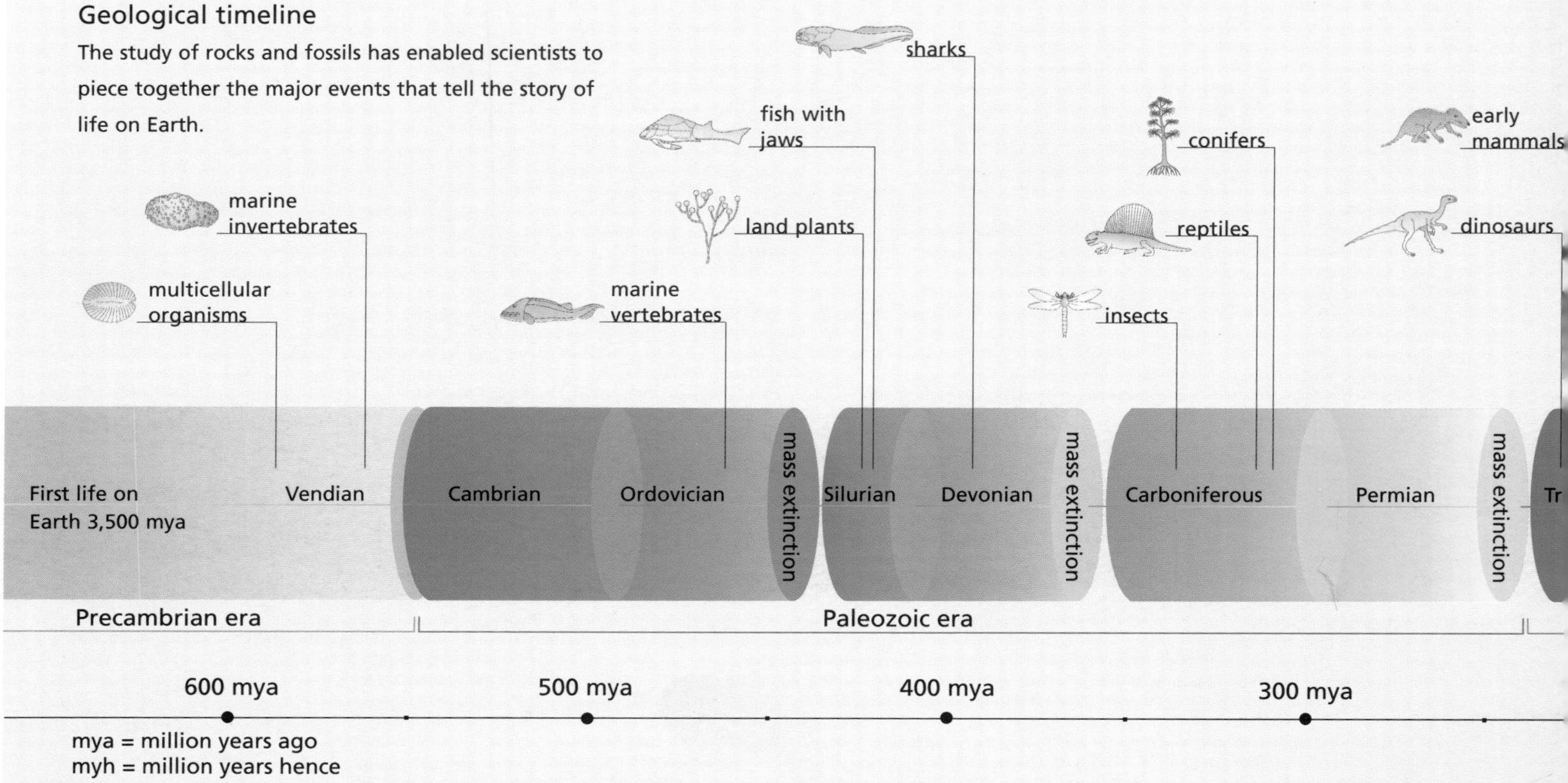

sharks

fish with jaws

conifers

early mammals

marine invertebrates

land plants

reptiles

dinosaurs

multicellular organisms

marine vertebrates

insects

| | | | mass extinction | | | mass extinction | | | mass extinction | |
First life on Earth 3,500 mya | Vendian | Cambrian | Ordovician | | Silurian | Devonian | | Carboniferous | Permian | | Tr

Precambrian era

Paleozoic era

600 mya   500 mya   400 mya   300 mya

mya = million years ago
myh = million years hence

Precambrian era

The Precambrian era is the longest era of the geological timescale, lasting around 4,000 million years. During the Precambrian era, the first simple bacteria appeared in the oceans. The presence of life caused oxygen to build up in the atmosphere, allowing more complex multicellular organisms and soft-bodied marine invertebrates to evolve.

Paleozoic era

Multicellular organisms underwent a dramatic diversification. Primitive plants and insects colonized land, while fish-like vertebrates thrived in the oceans. Sharks and fish dominated the seas and newly-evolving conifers spread rapidly on land. At the end of the Paleozoic era, the largest mass extinction in history destroyed 90 percent of all marine species.

Mesozoic era

There was an explosion of new life forms during the Mesozoic era. Dinosaurs became the predominant animals on land. The first flowering plants and mammals appeared and marine life continued to thrive. A mass extinction, most likely as a result of a meteor strike, brought the era to a close. Land-dwelling dinosaurs and many other species became extinct.

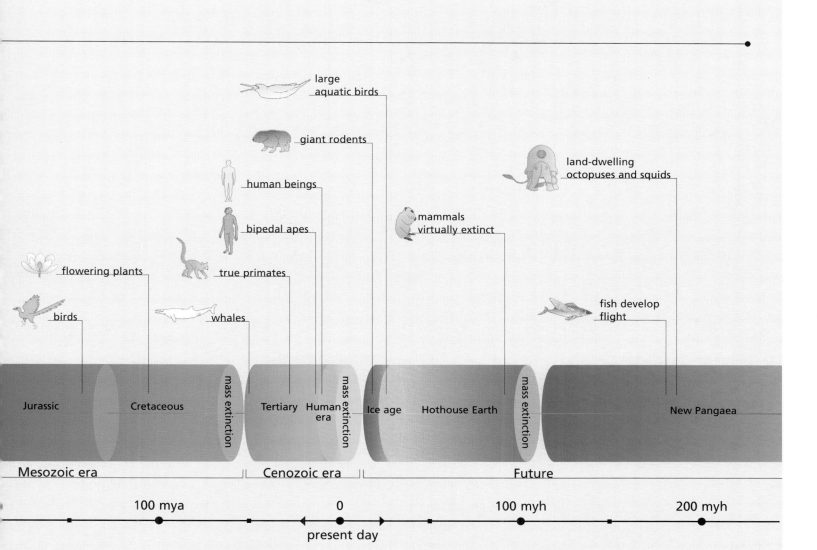

large
aquatic birds

giant rodents

human beings

land-dwelling
octopuses and squids

bipedal apes

mammals
virtually extinct

flowering plants

true primates

fish develop
flight

birds

whales

Jurassic

Cretaceous

mass extinction

Tertiary Human
era

mass extinction

Ice age

Hothouse Earth

mass extinction

New Pangaea

Mesozoic era

Cenozoic era

Future

100 mya

0

100 myh

200 myh

present day

Cenozoic era

Life flourished in the humid climate
of the early Cenozoic era. Flowering
plants, mammals and insects all
diversified on land. In the oceans,
the first marine mammals, early
whales, evolved. Midway through
the Cenozoic era, the climate became
drier and cooler, signaling the start of
a long-running Ice age. Hominids, the
ancestors of humans, appeared.

Human era

This book refers to the age of humans
as the Human era. Our brief existence,
which begins and ends in the latter
half of the Cenozoic era, covers a
short period of geological time, but
has a devastating impact on the
planet. The Human era ends a few
thousand years from the present with
a huge drop in temperature as the
Ice age reaches its peak.

Future time

Few species survive the Ice age that
led to the extinction of humans, and
successive mass extinctions in the
future wipe out large numbers of
land and marine species, including
mammals, fish and many birds. But
evolution responds in surprising ways.
Fish abandon the oceans and take to
the air, while octopuses and squids
become the dominant species on land.

5
MILLION
YEARS

ICE AGE

"Ice ages are huge perturbations to the natural system. As Earth becomes covered by ice, life is compressed towards the equator, and Earth's habitable area is vastly reduced. The climate changes immensely, so that organisms find themselves driven to extinction simply because it gets too cold and there's no place they can retreat to. An Ice age is a major transition in the history of life."

Professor Bruce Tiffney
Paleobotanist
University of California

FIVE MILLION YEARS HAVE PASSED since humans lived on Earth, a relatively short period of time in geological terms. The continents have drifted slightly, but not by much. Plants and animals have evolved and adapted, but they still have a lot in common with their Human-era relatives. The big difference between this time and the Human era is the climate. Earth is currently at the peak of an Ice age that has been ongoing for around seven million years.

When global temperatures drop, ice sheets advance outwards from the poles and down from the mountains. The glacial cycle of an Ice age lasts somewhere in the order of 100,000 years. Of this period, perhaps 90,000 years will consist of a cold spell, whilst 10,000 years will consist of a warm spell, known as an interglacial. The Human era took place during one of these warm interglacial periods, a time when the ice sheets had retreated.

Now, five million years after the Human era, icecaps have crept across much of the northern hemisphere. Much of North America is under ice. Where the ice sheet ends, midway down the continent, lies the arid, freezing expanse of the North American Desert. In what was once Europe, the story is much the same. Ice domes cover the whole of Scandinavia, and much of the rest of Northern Europe has become a broad tundra habitat of permanently frozen subsoil and sparse vegetation.

Between the two polar icecaps, in the more temperate zones of Earth, life is every bit as harsh. There is so little moisture in the atmosphere that even the dense, lush rainforests of the Amazon have been reduced to dry, windswept grasslands. So much water is locked into icecaps that the global sea level is about 500 feet (150 meters) lower than it was during the Human era, isolating the Mediterranean Basin and turning it into a region of parched salt flats.

The Human era ended with a period of mass extinction resulting from a combination of human influence and natural phenomena. Humankind's energy consumption had far-reaching and devastating effects. Ecosystems were fragmented and habitats destroyed. Then, with the last glacial advance, the situation worsened. The spreading ice sheets and falling sea levels eliminated whole ecosystems, leading to the demise of thousands of species.

The only animals to survive such a boom and bust cycle of glacials and interglacials were the generalists – animals able to exist under virtually any conditions and in different environments. Now, five million years after the Human era, these creatures have succeeded in colonizing the most inhospitable reaches of the planet.

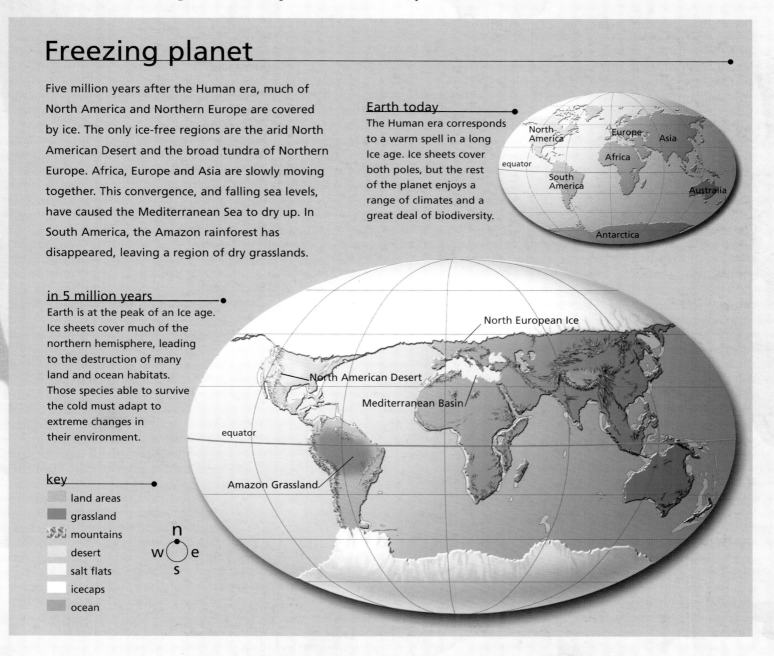

Freezing planet

Five million years after the Human era, much of North America and Northern Europe are covered by ice. The only ice-free regions are the arid North American Desert and the broad tundra of Northern Europe. Africa, Europe and Asia are slowly moving together. This convergence, and falling sea levels, have caused the Mediterranean Sea to dry up. In South America, the Amazon rainforest has disappeared, leaving a region of dry grasslands.

Earth today
The Human era corresponds to a warm spell in a long Ice age. Ice sheets cover both poles, but the rest of the planet enjoys a range of climates and a great deal of biodiversity.

North America

Europe   Asia

equator   Africa

South America

Australia

Antarctica

in 5 million years
Earth is at the peak of an Ice age. Ice sheets cover much of the northern hemisphere, leading to the destruction of many land and ocean habitats. Those species able to survive the cold must adapt to extreme changes in their environment.

North European Ice

North American Desert

Mediterranean Basin

equator

Amazon Grassland

key
land areas
grassland
mountains
desert
salt flats
icecaps
ocean

n
w   e
s

25

the NORTH EUROPEAN ICE

THE WORLD IS DEEP IN A PERIOD OF GLACIATION. Five million years after humankind, the planet is once again dominated by icecaps, as it was in the Pleistocene epoch, two million years before humans. So much water is locked up in the icecaps that global sea levels are nearly 500 feet (about 150 meters) lower than they were during the Human era. Ice sheets cover most of North America and the whole of Scandinavia, reaching down into Northern Europe.

Where once the craggy coastline of southwest England boasted semi-tropical plants which flourished in coves warmed by the Gulf Stream, there is now blizzard-swept tundra. The continental shelf which spreads out from the British Isles and France is an exposed, frozen plain of sand and gravel deposits – outwash from the glaciers – and frozen soil. On a winter's night, temperatures fall below -76°F (-60°C) and wind-chill makes it feel even colder.

Come the brief summer, conditions improve a little. The edge of the icecap lies sparkling to the north, its meltwater running in gravel-choked streams across the flat tundra. Water collects everywhere in lakes and ponds, unable to drain away through the permanently-frozen subsoil, or permafrost. Rocks lie in polygonal patterns, covering broad expanses of land. The bitter, boulder-cracking frosts of winter have heaved them up from the soil in successive freeze-thaw cycles and rolled them together in curving lines. Dome-shaped mounds, called pingos, rise up from the tundra. Formed around a core of ice, a pingo occurs where deeper water has sprung through the permafrost or in sites where lakes have progressively frozen from the sides towards the center.

Despite the harsh cycle of fierce winters and brief summers, there is life here. The permafrost does not favor deep roots, but certain forms of flora are tough enough to eke out an existence from the frozen soil. Clusters of cotton grass border the lakes, undulating meadows of small, hardy lichens and grasses cover the raised land between the gravel deposits and stream beds. Tight clumps of heather form a rooting site for many varieties of small flowers. The only plants which might be considered trees are closer to Human-era shrubs, and even these hug the ground. Species of willow send their gnarled trunks horizontally and spread their branches across Earth, as if unwilling to raise their heads into the bitter winds that will inevitably come with winter.

"In five million years time the Earth will be experiencing one of the biggest Ice ages in its history. Worst affected will be North America and Europe, where there will be ice domes up to two miles (three kilometers) high covering much of the USA, the whole of Canada and Scandinavia, and stretching down into Northern Europe."

Professor Paul Valdes
Paleoclimatologist
Reading University, UK

left

In Northern Europe, what little life has survived the Ice age must eke out a living on the frozen tundra.

left

Shagrats are the largest animals in the North European region. Their layered coats protect them from the cold climate of the northern tundra.

Spring has arrived in Northern Europe. Water from melting ice runs in narrow, muddy torrents between the banks of shingle that choke the river bed. The thaw softens the surface soil and turns large areas of the tundra into marsh, dotted with pools. Clouds of flies congregate above the pools. Because of the climate conditions, the fly life cycle is accelerated: they mate and reproduce in an incredibly short period of time during the brief thaw. Migrant birds wheel in and out of the swarm, making the best of the bonanza. Migrant birds fare particularly well here. They weather out the winters in less extreme regions further south and travel to these bleak lands in the fleeting summer months to take advantage of the quick and intense growth cycle of the flies.

A watery sun slants through the clouds of insects, illuminating the red fireweed, the yellow, powdery male flowers of the low willows and the white heather bells. All these plants grow in shallow pockets of soil, in contrast to the orange and yellow lichens that coat the frost-shattered rocks. A long, gentle slope of land runs down to a wide river bed, where the sunlight catches the water. The spring growth is well under way.

A group of animals is making its way across the network of streams and rocky debris. They are the size of sheep, although, with their ragged fur, they resemble the musk oxen that grazed the tundra regions of the Human era. Unlike musk oxen, these beasts have no horns. In place of hooves, they have broad, clawed feet. They are shagrats, the largest mammals of the northern tundra.

Despite their size, shagrats are rodents. They are descended from marmots, small burrowing rodents common in Europe, Asia and North America during human times. Of all mammals, rodents were in the best position to survive the mass extinctions that occurred towards the end of the Human era. Small and versatile, they were able to adapt to their changing environment. Indeed, it was this adaptability that led to rodents being thought of as pests in the past.

With the destruction of habitats and ecosystems and the extinction of so many plants and animals, rodents were able to step in and fill the ecological niches left vacant. Five million years after humankind, shagrats have developed a shaggy coat as a response to the harsh conditions. They are also herd animals, often huddling together for warmth. Otherwise, they resemble their smaller rodent ancestors, with large teeth and rounded cheek pouches. Such similarities are to be expected: in terms of evolution, five million years is a very short period of time.

The shagrat herd has spent the winter at the northern edge of the birch and conifer forest away to the south, where Paris was during the Human era. Now they are returning to their summer grazing grounds. The females give birth in early spring, following a gestation period which lasts throughout the winter. It is now the middle of spring, and about a third of the herd consists of youngsters.

The shagrats crunch through the banks of shingle and splash through the muddy torrents, their broad feet preventing them from sinking in. Their fur is so thick that a brief immersion does them no harm – they shake off the water before it soaks into the skin and chills them. Shagrat fur is layered as well as thick. An outside coat of coarse hair provides external protection, and a tightly-packed inner layer acts as an excellent insulator.

The herd scrambles up the bank at the far side of the stream and the shagrats begin to root about for food, pushing rocks over with their strong forelegs to reveal the soil beneath. The big claws on their forefeet dig into the soil to reach underground grass stems and the roots of heather and willow. Their marmot ancestors were burrowing animals and so their front feet were already well adapted for this kind of foraging. The shagrats spread out across the meadow. Claws scrape the ground, and plants are uprooted with broad chisel teeth. Suddenly, one of the shagrats barks a warning and the whole herd leaps to attention.

The tundra is broad and flat. Danger can be seen coming from far away. There are few dangerous animals here, but the big male shagrat, the leader of the herd, barks again. He senses a presence.

Surviving in the cold

Mammals have evolved several strategies to survive in a cold climate. A large body holds heat better than a small one. This is because large bodies have a relatively small surface area for the mass, so less body heat is radiated out. For this reason, many cold climate animals are particularly large compared with relatives in warmer climates. The nose, ears and eyes of cold climate animals are usually smaller, making them less prone to frostbite. Thick hair provides insulation. As a rule, dark colors are good for absorbing warmth when the sun is out. However, there can be a trade-off in this, as light colors provide better camouflage in snowy habitats.

In the Human era, the largest member of the deer family was the moose, which reached its maximum size in the cold climate of Alaska.

The herd is unsettled. The shagrats assemble in an open space and, instinctively, the youngsters huddle into the middle. The adults surround them in a defensive formation, all facing outwards. This behavior is very like that of musk oxen, which employed an identical strategy when under threat from wolves. Sure enough, there is a movement in the heather on the opposite side of the stream. A flash of white from a sleek coat betrays the approach of a stealthy predator. It is a snowstalker, the chief enemy of the shagrat.

The snowstalker is a mustelid, a member of the group of carnivores that once included wolverines, weasels and stoats. The snowstalker is much larger than any mustelid of the Human era, but is otherwise similar in most respects. One big difference, however, is its method of killing prey. The snowstalker's canine teeth have developed into long slashing weapons, designed for inflicting deep wounds on large prey. Evolution has witnessed such developments before. The snowstalker is essentially a saber-toothed wolverine.

Mustelids have always been solitary animals, never hunting in packs or family groups. The snowstalker is no different from its ancestors in this respect, and remains a lone hunter. It is territorial, but as the tundra on which it lives supports very few large plant-eaters, there is little in the way of regular prey here. The territory of a single snowstalker covers several square miles.

Because of the large distances between the territories of neighbors and potential partners, snowstalkers mate infrequently. Thin populations such as this are prone to inbreeding and the genetic weaknesses it brings, so the snowstalker has evolved a way to overcome the risk. The female snowstalker enters estrus, her fertile period, every 21 days. Each time she mates, the fertilized embryo is held in a kind of suspension. The embryos are eventually implanted in the womb and the litter is born as the winter ends and the summer thaw begins. This means that the young come from many different paternal lines. Delaying implantation in this way aids the spread of diverse genes.

left
A snowstalker patrols the rocky regions at the edge of the tundra. Its white pelt provides excellent camouflage in the snowy uplands.

left

Young snowstalkers play tug-of-war with a scrap of meat brought by their mother. The litter is born at the end of winter. This gives the pups a greater chance of survival as they grow up in the brief summer thaw.

Large prey is hard to find on the tundra, so when a snowstalker spots a herd of shagrats it will follow it for days. Finding food is particularly important during the summer thaw since, like the shagrats, the snowstalker has young to feed. Each day takes her further away from the lair where her litter is waiting, bringing her deeper into the territory of other snowstalkers.

The female snowstalker is at a disadvantage in the boggy plains of the tundra. Her chances of a successful hunt would have been better if she had waited until the shagrat herd was further up among the granite crags. There is still snow on the rocks, and the snowstalker's white pelt would have blended perfectly with the background. Normally she would make her move in a snowy terrain, but now she needs to cross an open stream in order to reach the herd. She must act quickly before the herd disperses. She creeps down the bank and across the stream, attempting to conceal herself in the heather on the other side. Too late. The shagrats are now well aware of the snowstalker's movements. The element of surprise is lost.

She emerges from the low cover and faces the herd. They bunch up even more closely, narrowing their eyes and baring their teeth, hissing in threat. She circles the mass of shaggy fur and threatening teeth, looking for a weak point in the protective formation, but there is none. She slinks back behind cover.

After a while the shagrat herd moves on. The snowstalker has lost interest in the attack. She is deep into the territory of another of her kind, a big male, and her hunting strategy must be adjusted accordingly. She will conserve her energy and let the male do the killing. The time will come when the male will ambush the herd, probably picking out a weak or inexperienced individual. He will dispatch the prey with his saber-teeth and monopolize the carcass, defending it against any other male that approaches. A wandering female, on the other hand, will be allowed to eat her fill in return for mating rights. Afterwards, the newly-inseminated female will return to her lair and regurgitate most of what she has eaten for her cubs.

Along the western coast of what was once France, the bleak tundra turns into an even bleaker shoreline. With its broad grey beaches of damp shingle, choppy slate-colored waters slicked with a mush of ice crystals, and huge icebergs, it is an inhospitable place to make a living. In the Human era, seals and sealions would have been virtually the only inhabitants of such an environment, but sea mammals have long since died out.

Where an ecological niche has been left vacant by extinction, something soon evolves to fill it. Five million years after humankind, the place of aquatic mammals has been taken by birds. Gannetwhales are similar in size and shape to a male walrus but are descended from gannets, large sea birds of the Human era. Gannetwhales roost on land and hunt fish underwater. Because of their large size they have given up flight, their wings evolving into stubby paddles, ideal for moving through water at speeds of up to 18 miles (30 kilometers) per hour.

Further physical developments allow the gannetwhale to hunt even more effectively underwater. As it dives into the freezing ocean, its nostrils close up to prevent it breathing in water. The gannetwhale is insulated from the intense cold by a dense coat of feathers and a layer of blubber, which also serve to streamline its body. The bird's diet of fish means that it ingests a lot of salt, which is excreted through glands above the eyes.

Like many marine birds of the Human era, gannetwhales have a tightly-knit family structure. The female lays only one egg at a time and tends it with great care. The mother insulates the egg from the bitter cold by clutching it to the underside of her tail with her feet, holding it where it will benefit most from the warmth of her body. During the incubation period – once the chick has hatched and is being tended by the female – the male is away fishing to bring back food for the whole family.

While the gannetwhale enjoys complete freedom from predators in the sea, the females and their young are at risk on land. To protect themselves from marauding snowstalkers, the birds nest on islands. But if a winter is cold enough, ice bridges will allow predators access to the colonies. Such conditions invariably give rise to the decimation of gannetwhale populations.

Return to the sea

In the Human era, gannets lived in large colonies on sea cliffs in the northern hemisphere. Like most sea birds, they spent much of their time chasing fish. However, gannets were unusual in that they could swim underwater with their wings. Since water is 800 times denser than air, moving effectively in both environments is difficult, and gannets had to compromise in order to be good swimmers as well as good flyers. Five million years after humankind, gannetwhales have adapted to an aquatic lifestyle by gradually losing their ability to fly.

A Human-era gannet plunges spectacularly into the sea in pursuit of fish.

right
Female gannetwhales guard their eggs from a hungry snowstalker.

the MEDITERRANEAN BASIN

MOVING SOUTH, AWAY FROM THE EDGE of the European ice sheet, the tundra stretches about halfway down to what used to be France. There, the bleak landscape is gradually replaced by clusters of trees such as rowan and birch. Further south, trees become more abundant and conifers appear. Isolated clusters become unbroken forests of conifers. Where the Alps rise, another band of glaciation occurs – towering mountains of ice creeping slowly down the valleys. Beyond these Alpine glaciers, stretching southwards, lies the Mediterranean Basin.

The Mediterranean Basin is a vast depression in Earth's crust, some 6,500 feet (2,000 meters) below sea level in places. It is now a region of brine lakes and salt flats, surrounded by dry, ridged limestone landforms, called karst. The air is dusty and parched. This is what remains of the gentle beaches, warm waters and hospitable climates of the former Mediterranean. The sea, around which great human civilizations once thrived, has dried up.

During the Human era, the Mediterranean Sea was fed by a constant influx of water through the Straits of Gibraltar. As global temperatures dropped, the icecaps expanded, and sea levels fell. The threshold of the Mediterranean became exposed and the flow of water stopped. This, coupled with the slow collision of the African and European plates, has left the Mediterranean landlocked.

Despite influxes from rivers flowing from Europe and Africa, the sea gradually evaporated. As the sea level dropped, the concentration of minerals in the remaining water rose, and vast limestone deposits began to form on the floor of the shrinking sea. As the water evaporated further, stretches of limestone pavement became exposed. It took about one million years for most of the sea to evaporate and now, five million years after humankind, any water left has collected in hypersaline lagoons in the low basins.

Out of the brine lakes and fissured limestone karsts of the Mediterranean Basin rise majestic mountains. These are the former holiday islands of Majorca, Crete and Cyprus, among others, that now stand tall and exposed on the harsh, dry plains. Global temperatures are five or six degrees lower than they were in the Human era, and nowhere is this difference felt more than in the Mediterranean Basin. This once warm, sun-drenched region is now an arid, cold and rocky land.

"In five million years, as Africa continues to jostle with Europe, the narrow Straits of Gibraltar, which are the only connection with the world's oceans, will be closed off, and so the Mediterranean Sea will become isolated. As a result, the water in the Mediterranean will gradually evaporate and we will be left with dry land."

Professor Paul Valdes
Paleoclimatologist
Reading University, UK

left

Five million years after the Human era, the Mediterranean Sea has dried up. The warm waters and hot sandy beaches are gone, leaving a region of salt flats and brine lakes.

35

The salt flats of the Mediterranean Basin shimmer in the cold sunlight. What little water there is lies in the deeper basins, forming scattered, shallow lakes. The waters in these lakes are ten times more salty than seawater. They contain no fish. The only life that can survive in such a hypersaline environment are simple algae and bacteria that feed on the rich chemical soup. At the edges of the salt lakes, large clouds of brine flies gather. These tiny black flies have always flourished under such conditions, and continue to do so now, feasting on the algae and bacteria in the shallows.

The surface of the Mediterranean salt plain is completely flat. On the rare occasions that it rains here, the top surface of the salt is dissolved into a saline mush. As the water evaporates, the surface is once again flattened. The whiteness is only alleviated by the odd smudge of red produced by the salt-loving bacteria. But there is something moving on the dazzling salt flat. It is a cryptile lizard, a member of the agamid family of lizards which has populated the dry regions of Earth since well before human times.

Now the lizard is hunting. It scampers along and raises itself on to its hind legs. Then it opens up a frill around its neck. Superficially, it resembles the frilled lizard, an agamid that lived in Australia during the Human era. But unlike the frilled lizard, which had an unbroken membrane, the cryptile's frill forms a net which is covered in a film of sticky mucous secreted by pores on the individual strands.

The cryptile runs across the salt, the open weave of its frill causing little air resistance. It careers headlong into a thick cloud of brine flies, running straight through it. The cloud swirls and parts and the lizard settles, exhausted after its brief bout of activity. The frill that was pale-colored before the cryptile's run, is now black and laden. The brine flies have adhered to it as if to flypaper. Settling in the salt, the lizard folds its frill, gathering the trapped insects into bunches. In rapid jerks it extends its long tongue and picks off the flies.

The cryptile never drinks water from the lakes – the amount of salt they contain would prove fatal if ingested. The lizard gains all the moisture it needs from the flies it eats.

Agama lizards

Human-era agamids, such as the thorny devil and the agama lizard, used their serrated skin and amazing displays of color to camouflage themselves in their natural habitat. A distant cousin of the cryptile is the frilled lizard, which also had an expandable neck frill. The frilled lizard activated its neck frill as a defense against predators, as a mating display or even as a heat regulator. Cryptiles have adapted their neck frills for a further function: to net the brine flies which are their main food source. As the cryptile runs through a cloud of insects, the flies stick to its frill and the lizard can pluck them off at its leisure.

A Human-era frilled lizard spreads its neck frill to frighten off predators.

left

A cryptile runs headlong into a cloud of brine flies, with its neck frill spread wide. Brine flies provide the cryptile with all the protein and moisture it needs.

overleaf

A pair of cryptiles dash across the salt plain in an energetic mating ritual. The male has attracted its mate with an impressive display of color.

After hunting, the cryptile's pale coloring allows it to fade back into the white of the salt flats. It spreads its ribs and flattens its body against the salty surface. The rough, three-dimensional pattern of its scaly skin mimics the crystalline salt structures of its surroundings. So positioned, the animal is practically invisible. Indeed, the cryptile is a true master of disguise and is well-deserving of its name, which is derived from the Greek word kruptos, meaning hidden or concealed.

There are times when the cryptile lizard abandons its camouflage and changes its color to deliberately draw attention to itself. When an enemy approaches, the lizard quickly adopts a threatening pose. It puffs itself up and spreads its frill, turning black in color. Against the white background, the cryptile now stands out, appearing to be a bigger and much more threatening creature. This is usually enough to frighten off an attacker, and allows the lizard to make its escape.

During the mating season, the cryptile adopts yet another colorful appearance. As a prelude to mating, the male cryptile raises itself with its front legs and expands its frill. Cartilaginous ribs allow the frill to open rigidly like a fan or an umbrella. Pigment cells in the frill flash vivid colors across the structure. The more impressive the display, the greater the chance of attracting a potential mate. This skin coloration also serves as a warning signal to rival males. At such times the cryptile stands on top of salt pinnacles or other promontories and allows itself to be seen from afar.

Female cryptiles also have an expandable frill but, unlike males, do not use it for mating displays. Instead, the frill is only used for feeding and in self-defense. As soon as she spots a displaying male, the female can cover great distances across the salt plains to meet her mate. Once mating is over, the female leaves the salt lagoon and heads into the rocky limestone karst in search of a safe place to lay her eggs.

Around the edge of the salt lagoon, the landscape gradually changes from white to grey. This is the karst – a region of rocky limestone blocks and pinnacles, separated by deep fissures called grykes. Grykes are formed by the gradual action of rainwater on natural cracks and faults in the rock. Over time, the water has a weathering effect on the rock, dissolving the limestone and deepening the fault lines into larger fissures until, finally, they become grykes. What soil there is in this dry landscape gathers at the bottom of the grykes, enabling some vegetation to take root.

It is in the shelter of the grykes that the cryptile hides her eggs. They will be safer here than on the open salt plains, but there are dangers nonetheless. Snuffling about in the broad passageway of a gryke comes a scrofa, a descendant of the wild boars that once ranged across Europe and Asia. Human-era wild boars were hardy creatures, adaptable to changing conditions and generalized in their feeding habits. They fed largely on grain and roots, but occasionally would kill and eat small animals.

Five million years after the Human era, scrofas are similar in many ways to their ancestors, although they are smaller and more lightly-built. Delicate hooves and agile limbs allow them to prance along the wide slabs of limestone pavement, known as clints, leaping over the grykes as they go.

The generalized, omnivorous diet has not changed much either. Scrofas will eat almost anything. Plant material forms a large part of their diet, but they will also prey on slow-moving animals and carrion. Their cylindrical nose, used for searching out food from the bottom of grykes, is longer than that of their wild boar ancestors, and is now a short, flexible trunk.

The long nose of a snuffling, foraging scrofa finds the newly-laid cryptile eggs in the shallow soil. Its tusks loosen the eggs from their hiding place and its long lips snap them up, then it moves on. This was one little treasure the scrofa had no intention of sharing with other members of the herd grazing nearby.

left

A herd of scrofas forages in the grykes for vegetation and even small animals. The scrofa's long snout allows it to reach into the deepest crevices.

left

A gryken pokes its head out of a gryke. It is tracking a herd of scrofas and their young, called scroflets.

The scrofas spread out across the karst. There are two dominant females – the joint leaders of the herd. A scrofa herd is essentially a family group, consisting of the leaders and their offspring. The only males in the group are juveniles. Once the males reach maturity, they leave the herd to begin hunting and foraging on their own, mating with the dominant sows of other herds and fighting with other lone males for mating privileges.

In the late spring, scroflets are born in litters of between three and six. This is the best time of year for raising young, as there is more food available in the bleak environment. Wherever scrofas go foraging, there are always two or three on the lookout for danger. Birds of prey circle in the skies above the karst, but other dangers lurk in the grykes themselves.

Down by the edge of the salt flats, the scrofa herd is relatively safe – but there is little food to be found there. On slightly higher ground, soil particles blown into the grykes from the continental interior allow vegetation to grow. Patches of wispy grass sprout up from the flat surfaces of the clints. Hazel trees grow in bunches along the grykes, stunted and gnarled, their roots reaching down hundreds of feet to whatever water table exists here. This is the best feeding ground for scrofa herds at this time of year, since vegetation is relatively abundant. Unfortunately, it is also the domain of the scrofas' greatest enemy.

Down in a narrow gryke, a slim, agile creature winds from side to side, following the tortuous route of the chasm. It is a gryken, and it is tracking scrofas. Now and again the gryken stops and pokes its head out of the gryke, keeping an eye on the scrofa herd. Then its head is down again and it continues to negotiate the labyrinth of fissures that separates the clints. The chasms tend to run at right angles to one another, a legacy of Earth-moving forces that cracked the rocks originally. In places, the gryken's path is interrupted by a gaping hole, an entrance to the network of subterranean caverns that lies below the karst.

41

The slinky predator senses that it is close to the scrofa herd. It pauses near a stand of hazel trees. Silently, it stands on its hind legs between the stubby trunks and peers out through the leaves. Its triangular face is crossed by a thick black band which helps to camouflage its beady eyes. The sleek body is also striped, breaking up the gryken's contours and helping it to blend in with the vegetation. A litter of four scroflets is rooting around in the straggly grass of the surface, and one is just a little too far from its mother.

The hunter is poised to leap from the gryke and streak towards the hapless infant. It is a fast, streamlined predator. Like the snowstalker of the cold northern plains, the gryken is a member of the mustelid family. Descended from the tree-living martens of the Human era, it has the same long fur-coated body and triangular head. Its ancestors had powerful hind legs which they used to leap through trees, and a long tail to maintain balance as they ran along branches. The gryken, on the other hand, has had to adapt to a habitat of rocky crevices. It no longer needs a long tail for balancing and its legs are long and slim – better suited to squeezing through narrow cracks.

The stealthy gryken's jaws are full of sharp teeth, now bared for the kill. Its canines are much longer than those of its ancestors, reaching down over its lower jaw. Like the snowstalker of the North European tundra, the gryken is almost saber-toothed. Its long teeth are needed to tear through the tough hides of scroflets, the gryken's main food source. The victim is swiftly disabled, its throat punctured and its windpipe ripped out.

The scroflet has not seen the danger and the swift attacker sinks its sharp canines into the baby's throat. Too late, the scroflet lets out a plaintive squeal. Its mother looks up and, with an enraged scream, leaps over the grykes, displaying her deadly tusks in attack. Her attempted rescue is in vain. The gryken has thrown its prey down into the cave mouth. In a flash, it is gone.

Mustelids

Mustelids are a family of solitary, carnivorous mammals. Human-era mustelids included land-living species, such as badgers, weasels and stoats; aquatic species, such as otters; as well as tree-dwelling species, such as martens. The pine marten, from which the gryken has evolved, was widely exploited for its thick brown fur during human times. Pine martens lived in the hollows of trees and hunted rodents, birds and eggs high in the tree-tops, leaping from branch to branch with their strong limbs and using their tails for balance. As global temperatures dropped, forests declined and the pine marten's habitat began to disappear. It was in response to these changing conditions that the land-dwelling gryken evolved.

Pine martens were found in the wooded regions of Europe and Asia during the Human era.

right
A female scrofa's attempts to rescue her baby from the gryken are in vain. She will not be able to catch the predator as it leaps into the gryke.

the AMAZON GRASSLAND

ICECAPS MAY DOMINATE THE GLOBE five million years after the Human era but, between the equator and the poles, there is still a diverse range of habitats. There are equatorial rainforests, tropical grasslands, tropical deserts, temperate woodlands, coniferous forests and tundra. Over time, these habitats have been compressed into narrower and narrower bands by the cooler climate.

During the Human era, the actions of humans threatened to destroy the equatorial rainforest. Five million years after humankind, it has been further depleted by natural changes brought about by the Ice age. In the Amazon Basin, the forest is reduced to scattered pockets, surrounded by broad stretches of savannah which reach out to the horizon. Rainfall is lower now than it has been for millions of years and the mighty Amazon river has dwindled. The vast network of tributaries that once made it the most voluminous river in the world have all but dried up.

The dry conditions mean that the savannah is frequently swept by bushfires. These can be triggered by lightning strikes or a glint of sunlight. Once ignited, the fires can sweep across hundreds of square miles at great speeds, destroying everything in their path. In the Amazon Grassland, life has adapted to drought and fire in surprising ways. The grasses rely on the fires to clear the ground of other competing plant species, such as trees. If left to grow unchecked, trees can dominate the land, cutting out the sun with their thick canopies and monopolizing the soil with their roots. Grasses, on the other hand, are fast growing plants and easily replace their burnt leaves with new growth from underground stems. They disperse their seeds when the fire has passed, and thrive in the dry habitat.

The forests that once occupied this region were far more productive than the Amazon Grasslands. But as the forests declined, so too did the number and diversity of animal species. Because South America is connected to the rest of the world only by the narrow strip of Central America, a connection which is periodically broken by changing sea levels and volcanic activity, animal life has developed in the savannah in virtual isolation. Cut off from the rest of the world and faced with drastic environmental changes, those species which survived extinction have evolved new adaptations for life in this ecosystem. The animals here have lost the shelter of trees and are now exposed to wind and fire. Where once food was plentiful, animals must now cover great distances to find it.

"During the next Ice age, the global mean temperature will fall by only a few degrees. However, the effect of this drop in temperature will be very pronounced in areas such as the Amazon Basin where the climate will become much drier. The tropical rainforests we see today will be reduced to a few small clusters and replaced by large areas of dry savannah grassland."

Dr Roy Livermore
Paleogeographer
British Antarctic Survey, UK

left
The rainforests of the Amazon have all but disappeared, leaving a dry savannah. The region is frequently ravaged by swift, intense bushfires.

45

left

Babookaris are well adapted to life on the ground and can cover large distances in search of food. They communicate with other members of the troupe by waving their long tails above the grass-tops.

Out on the grassy expanse, a bunch of unusually tall, thick grass heads is waving in the breeze. Indeed, it seems that the grass heads are not just waving, but actually moving through the grass. These are the tails of a group of animals, about thirty strong. Occasionally, one stops and raises its head above the tops of the grass stems. The face is round and almost human, its skin naked and red, framed in a lion-like mane. These animals are monkeys.

The deep rainforest that once grew here was home to a number of monkey species. There were large monkeys that lived on the forest floor, agile monkeys that swung about in the branches and tiny, acrobatic monkeys that leapt from tree-top to tree-top. The fertile habitat could sustain such variety. Now the rainforest has gone, replaced by an unchanging stretch of grassland. As it shrank to islands of trees and, finally, to isolated stands, those species of monkey which could not adapt became extinct.

One species not only survived, but flourished. The uakari was one of the most adaptable of Human-era Amazonian monkeys. It was an omnivore, eating anything from insects and fruits to leaves, seeds and even small vertebrates. It was at home in the trees, but just as comfortable on the ground. Such generalism left it well positioned to adapt to the retreat of the rainforest. The uakaris were able to abandon their arboreal way of life and take to living on the grasslands which were spreading across the Amazon.

This is not the first time that rainforest has been replaced by grassland. Several million years before the Human era, a similar phenomenon occurred in Africa, brought on by a drop in atmospheric humidity levels. Then, the ancestors of baboons left the trees and took to dwelling on the ground, becoming more quadrupedal in the process. Now, in South America, the descendants of the uakaris have done exactly the same. They have evolved into babookaris.

Like all monkeys, the hands of a babookari are prehensile, that is, they are adapted for grasping or gripping. Now, though, they are more often used for walking than for swinging through trees. The babookari is essentially a larger version of its uakari ancestor, down to its hairless red face. One major difference is the babookari's tail. The uakari was the only South American monkey that did not possess a long, prehensile tail. Its descendant has evolved a long tail, but this is not a muscular, extra limb used for swinging about in trees. Instead it is a tall, inflexible rod with a plume of hair at the end. Longer than the deep grass, this tail is used for signaling across the plains.

Being a social animal, the babookari needs signaling devices such as its long tail and colored face. The only way a monkey can live on the open grassland is as part of a troupe. A large group of thirty or so individuals can quickly scour a wide area of savannah for food and can co-operate in defense against a predator.

Since it came down from the trees, this monkey has gained a great deal of intelligence. It has retained enough dexterity in its hands to be able to weave complex structures from grass stems, and has the knowledge to put these structures to work. One of the structures it builds is a fish trap, a hollow spherical basket which the babookari deploys in the shallow seasonal rivers that wind their way across the plain. Fish is an excellent protein supplement for these omnivorous monkeys. When on fishing trips, the troupe stays together. Some individuals work the traps, while the others are on the lookout for danger. On the open grassland, danger can come in many shapes and sizes. Almost certainly, however, it will come on long legs.

On a grassy plain, even if the grass is long, there is little cover. Large animals cannot hunt by stealth, so instead they must hunt by speed. Back in the Human era, the most famous of the grassland predators was the cheetah – the fastest animal on Earth at that time.

left

Babookaris are dexterous enough to weave fish traps from the long grasses. They place the traps in the rivers which wind across the savannah during the wet season.

The swiftest hunter of the Amazon Grasslands is a flightless bird called the carakiller. Flightless birds are not uncommon. Flying is a useful skill, but if there are few predators to escape from, and when there is enough food on the ground, birds often discard their powers of flight. Flightless birds have long been successful on grasslands, particularly those birds large enough to defy other predators. The Human era saw ostriches in Africa, emus in Australia and rheas in South America, all in grassland habitats. These ancient birds were all related, and it may be that they evolved from the same flightless ancestors. However, the history of bird evolution is full of stories of flying birds that evolved into flightless species. The dodo is just one example of a flightless bird which evolved from an ancient species of flying pigeon.

One of the most common falcons of the Amazon Basin during the Human era was the caracara. The caracara was a versatile, omnivorous bird, and easily adapted to life on the plains. Its descendants became large ground-dwelling birds and eventually evolved into a bird that was completely flightless – the carakiller.

The wings of the carakiller no longer have large flying feathers, but they are still muscular and aerodynamic. The bird's flightless wings now act as stabilizers, balancing the carakiller when it runs at full tilt, and helping it to turn corners quickly. In addition, each wing has a long curved claw at the tip.

The carakiller's body is covered in feathers for insulation, shaggy on the back and legs, and fine on the chest. Its neck and wings are bare, presenting a smooth surface when the bird is eating – feathers would soon become sticky with the flesh and blood of prey.

Carakillers live and hunt in loose groups, stalking across the grassland in open formation, looking for the telltale movements of a troupe of babookari. At about seven feet (over two meters) tall, the carakiller can easily see its prey over long distances. When a troupe of babookari is spotted, the carakillers signal silently to one another, raising and lowering colorful, peacock-like plumes on the backs of their heads.

Marabou stork

The Human-era marabou stork stood around five feet (1.5 meters) tall and had a wingspan of up to 9.4 feet (2.9 meters). Like the carakiller, the marabou stork ate virtually any animal matter, from insects to large game carrion. As an inhabitant of the African savannah, the marabou stork was witness to frequent fires and used them to its advantage. It would hunt ahead of the line of fire, picking out animals as they fled the flames. The carakiller also uses bush fires to hunt but, unlike the Marabou stork, it cannot fly away from the red hot flames. It relies purely on its long legs and swift gait to carry it away from the fire.

Human-era marabou storks were scavengers, finding food everywhere from lion kills to garbage heaps.

right
A female carakiller guards her eggs. Carakillers communicate by raising and lowering colorful head feathers.

The carakillers begin to close in on the monkeys. Suddenly, one of the babookari lookouts spots them and shrieks out a warning. They scatter, whooping and screaming, trying to confuse their attackers. The carakillers single out one of the babookaris and swiftly run it down. The other monkeys reform some distance away and carry on with their lives.

At times, carakillers employ a different hunting strategy, using the frequent bushfires to their advantage. As the fire races across the savannah, the animals of the grasslands run for their lives. Carakillers can do this easily, but other animals are not so swift. The birds run ahead of the flames, snapping up small mammals, snakes and lizards as they are flushed from their hiding places. Other birds walk behind the line of fire, picking at the charred corpses left there.

There is one animal that is supremely adapted to dealing with the periodic fires of the grasslands. It is a descendant of the the paca, a rodent indigenous to South America during the Human era, and it is called the rattleback.

The rattleback is about the size of a Human-era otter. It is covered by an armor of thick protective plates which give it the appearance of another Human-era creature – an armadillo. The scales which form the plates are actually great mats of hollow hairs. The air trapped inside the hairs provides insulation against heat. Beneath the roof of scales, there is a layer of insulating pelt, and the animal has rows of quills along its flanks. The hairy plates are the rattleback's first line of defense against bushfires.

In most cases hair is soft and fluffy, trapping air between its strands and acting as an insulator, to keep a mammal warm in cold climates and cool in hot climates. But hair can also be highly specialized. Porcupine quills were individual hairs, grown stiff, strong and pointed, while the horn of a rhinoceros was made entirely of compacted hair. The basic protein in hair cells is keratin. In fact, it is not just hair cells which contain keratin: skin, nails, feathers, beaks, horns and hooves are also based on this one substance.

left

Carakillers use the frequent grassland fires to their advantage, hunting along the fireline and snatching prey from the advancing flames.

left

Armored plates of thick, stiff hair make rattlebacks durable creatures, able to withstand bushfires and defend themselves from angry carakillers.

In the case of the rattleback, the keratin in its plates has mineralized – minerals from the animal's diet have combined with the keratin to make it harder. In a fire, the rattleback flattens its flameproof back plates, digging them into the soil around its body. Hunkered down in this way, it can weather out the firestorm as it sweeps overhead.

Fireproof plating would be of little use if the rattleback's eyes were not also protected, but it seems that this rodent has all bases covered. In a firestorm, it shields its eyes with a thick layer of hardened skin. When the fire has passed, the rattleback opens its eyes and surveys the damage. At worst, it will have suffered a scorched plate or two, and these are easily regrown. Of much greater interest are the charred animal corpses which litter the burnt ground.

Food is scarce on the Amazon Grasslands and so the rattleback is an opportunistic feeder. Its diet consists mainly of grass stems and buried tubers, which it digs out using its large front feet and strong claws. However, when a fire passes overhead, killing everything in its path, the rattleback is on hand to pick over the spoils. As the flames die down, the rodent helps itself to a barbecue of burnt carrion.

Another tasty protein supplement comes in the shape of carakiller eggs, laid on the ground during the wet season, when fires are less frequent. The rattleback shows no fear as it approaches the nest. It simply breaks open an egg and eats it on the spot. Should the enraged mother return to the nest, the rattleback flattens itself to the soil, wedging itself into the ground with its sharp quills. The carakiller's beak is useless against this armor and no amount of scratching and clawing will pry the rattleback loose.

Rattlebacks are solitary animals, coming together only to mate. Competition for food makes them highly territorial and they defend their foraging areas by clattering their back plates. The distinctive, aggressive noise warns off potential intruders. It is this rattle that gives the rattleback

51

the NORTH AMERICAN DESERT

THE NORTH AMERICAN CONTINENT IS COVERED by an ice sheet that ends just south of the border between what were once the United States and Canada. The central region of North America is cold and dry. The atmospheric temperature is so low that the air has little capacity for holding moisture. What was once the most productive agricultural land on the planet is now little more than a vast, barren dust bowl.

Along the eastern edge of the continent, the Atlantic Ocean has receded, leaving a broad coastal plain topped by the Appalachian Mountains. Inland, there is an unending expanse of cold sand and cracked rock. The North American Desert is as bitterly cold as the Gobi once was in central Asia. It stretches for about 1,500 miles (2,400 kilometers) until it finally fetches up against the rocky barrier formed by the glacier-ridden Rocky Mountains away to the west.

All year long, piercing winds sweep southwards from the ice, stirring up vicious sandstorms and scouring away any potentially productive pockets of soil. Occasionally, winds howl in from the coastal regions, but very little rain reaches the interior. What precipitation there is tends to be snow, but even this falls infrequently. Any snow cover on higher ground is thin and patchy. There is a significant difference in temperature between the equator and the edge of the icecap, producing a steep temperature gradient from north to south. This leads to unstable conditions with screaming tornadoes ripping across the interior far more frequently than they did during the Human era.

Vegetation is sparse in this harsh, arid landscape. Only the hardiest of plants exist, but in small numbers. Indeed, very little life can survive above ground. Those animals able to live in this bleak desert must withstand freezing temperatures and protect themselves from the violent storms. The animal inhabitants of the North American Desert are supreme specialists, brilliantly adapted to cope with the specific demands of an uncompromising habitat. Some species have adapted physically, whilst others have altered their living environment. Animals that live here need to last out times when food is in short supply. In order to do so, they have developed remarkable strategies for storing energy and conserving food. Some animals have even developed altruistic feeding habits, sharing food with other colony members to ensure the survival of the species as a whole.

"Five million years after the Human era, much of North America is under ice. South of the ice sheet, the rest of the continent is subject to extraordinary winds, desiccation and absence of moisture. It is a cold, dry desert. Plants find it very difficult to grow in this environment because of the cold, the lack of moisture and also the sand-blasting effect of the winds, which constantly shift loose sediment around."

Professor Bruce Tiffney
Paleobotanist
University of California

left

The fertile agricultural belt of North America has been frozen out of existence by the advancing ice. The rolling fields have been replaced by a freezing, featureless desert.

53

left

A desert rattleback trudges across the rocky earth. Its tough, hairy scales provide excellent insulation against the bitter desert winds.

Central America – the narrow strip of land, or isthmus, that links North and South America – has not always been a true land bridge. For many millions of years before its formation, the two continents were quite separate and distinct. When the connection first became permanent, land-dwelling mammals migrated from north to south, while only a few spread in the opposite direction into North America. By the time the Human era dawned, one of the few South American mammals to have flourished in North America was the opossum.

Now, five million years after the Human era, another southern mammal has made the journey – the rattleback. The rattleback evolved from the paca, a large burrowing rodent found in the forests of South America during the Human era. South American rattlebacks were highly successful in the dry grasslands that were spreading across the Amazon basin. So much so, in fact, that they were able to migrate northwards into the North American desert. There, the rattleback evolved into yet another species, one adapted for the cold desert environment.

The desert rattleback is a larger animal than its grassland cousin. It has evolved a large body in response to the cold climate. Big animals have less surface area relative to their body mass, making them more efficient at keeping in the heat. However, despite the desert rattleback's larger body size, its nose, ears and lips are much smaller than those of the grassland rattleback. In such a cold climate, the smaller an animal's extremities, the less susceptible they are to frostbite.

The hairs on the backs of both species of rattleback have evolved into hard, interlocking scales, although the desert rattleback's scales are smaller than those of the grassland rattleback. Because there are fewer predators in the desert, the scales need not be as strong as those of their grassland relatives. Heat insulation, on the other hand, is vital in the cold desert, and large air pockets within the scales provide excellent protection against the elements. The desert rattleback's face is covered in thick hair to shield its eyes and nostrils from the piercing, wind-borne sand.

Food and water are scarce in the North American Desert, so when a rattleback comes across food, it gluts itself. What nourishment is not immediately used will be stored away as fat reserves to see the animal through times of famine, in much the same way that Human-era camels stored fat in their humps. The rattleback's kidneys are also very efficient. It cannot afford to waste water, so its urine is highly concentrated. In fact, the rattleback hardly needs to drink at all. It obtains nearly all its moisture from the food it eats.

Unlike its omnivorous southern cousin, this rattleback is a plant-eater, subsisting mostly on the tubers of desert turnips. Its acute sense of smell can detect plants from the surface and it uses broad, clawed feet to dig them out.

Although the desert rattleback spends much of its time burrowing below the desert surface, it is not a subterranean creature. It does not dig tunnels and pits as other, more specialized, desert animals do. During a violent sandstorm, however, it works its way into the soft sand and dust with a kind of a swimming action, shuffling down with its broad feet. Then it lets the displaced sand spill back, partially burying itself.

One animal that avoids the grim surface conditions by living below the desert is the spink, a species of burrowing bird descended from the quail family. It has long since given up flying. Instead, its wings have become adapted to digging. On the underside of each wing, the feathers have become horny scales, forming an abrasive surface. The articulation of the spink's forelimbs has also changed radically. Instead of flapping in time as birds normally do, the limbs move independently, shoveling away at Earth. The spink crawls in much the same way as it digs, levering itself forward with its elbows, its weight supported by horny pads at the joints.

right
This strange burrowing bird is a spink, a distant cousin of the Human-era quail. The spink's wings have evolved into strong forelimbs for digging tunnels deep below the desert.

The rest of the spink's body is covered in fine black and white feathers, not unlike the downy feathers of baby penguins in the Human era. However, with their elongated bodies, spade-like forelimbs and strange, crawling gait, spinks bear little resemblance to Human-era birds.

Spinks live in large underground colonies, deep below the inhospitable desert surface. Digging such a vast network of tunnels requires a degree of organization, and the spinks work together. They form chain gangs. The birds first loosen the soil with their beaks and then scoop the soil behind them with their wings, passing it to one another along the tunnel. In their dark subterranean habitat, spinks do not need to see. Their eyes have been reduced to mere pinpricks, like the eyes of Human-era moles. They communicate by sound or touch, twittering and squeaking in the tunnels.

Nearly all individuals are biologically juvenile, having never reached the reproductive stage. Only one female member of the colony, a queen, will mate and lay eggs. She is able to deposit hormones within the egg that determine the sex of the baby and whether it will be able to breed or not. All the work of the colony revolves around keeping the queen alive and nurturing the brood.

The spink's diet consists of the same desert turnip favored by the rattleback. Spink colonies are established where turnips are most abundant. Due to the limited food supply, the few fertile males and females pair up to establish new colonies, traveling across the desert in search of turnip plants. Spinks are in great danger on the desert surface, and so they emerge from their tunnels only under cover of night. During the early mornings and late afternoons, hungry predators circle overhead.

left

A group of spinks gathers at an intersection between tunnels. Spinks communicate with a twittering, squeaking song.

left

On the ground, the deathgleaner supports its body by resting on its elbows. The helpless spinks will be easy pickings for this predator.

High above the shifting sands of the desert, black, winged creatures hang in the air, resting on updrafts of wind blown off the sandy ridges below. They circle like Human-era vultures, seeking an easy meal on the ground. Now and again, one wheels and banks, signaling to the others that it has found something. Soon, a large group has gathered and is preparing to feast.

The flying creatures are not vultures, but deathgleaners: predatory, scavenging bats with wingspans of four feet (1.3 meters). In this instance, it is not food they have spotted, but something on the ground that suggests food may not be far away. A deathgleaner's penetrating eye has noted a shuffling in the sand. It is a desert rattleback on the move, looking for nourishment of its own. On occasion, deathgleaners will take a young rattleback, but this is not common – the rodent's armor is tough and unpalatable. Instead, the bat will follow the movements of a rattleback in the hope that it leads to an easier and more abundant source of food.

Now the desert rattleback is among the thick, succulent fronds of some desert turnips, and is digging down to the tubers. Soaring deathgleaners circle in, waiting. Beneath the patch of turnips, there is a spink colony – a network of tunnels and chambers surrounding a substantial growth of tubers. The colony has been there since the turnips took root, and the ground is unstable.

The unsupported earth beneath the rattleback gives way, forming a crater around the plants. The rattleback sinks from sight and then surfaces again, spitting sand and scrambling to firmer ground. Immediately, the loose sand around the rattleback is alive with spinks, which attack the floundering intruder in a futile display of defense. Most of the spinks are quick to flee the ruined nest, but those in the collapsed tunnels closest to the surface cannot penetrate to the deeper, secure regions of the colony. In a state of panic, they crawl clumsily across the desert surface looking for shelter, but their efforts are hindered by their virtual blindness.

57

Down come the deathgleaners, shadow after shadow swooping in. Long talons, more bird-like than bat-like, pin the spinks into the sand. Strong jaws and enormous teeth crush into their backbones, delivering death swiftly. The shattered spink colony will provide plenty of food.

A deathgleaner's wing is typical of the wing of any bat. It consists of a membrane of skin stretched out between the elongated fingers of the hand. This is not the ideal arrangement for a cold-climate animal, since a great deal of body heat is lost through the skin. Birds do not have this problem because their wings are covered in insulating feathers. Deathgleaners avoid excessive heat loss by cooling the blood before it is pumped through the veins of the wing membrane. The heat taken from the blood going into the wings is used to warm the cooled blood coming back from them, by the same principle as industrial heat exchangers.

Deathgleaners live in communal roosts, in the caves and ravines of the distant Rocky Mountains. They sleep during the freezing desert night, huddling together to keep warm, and preserving enough energy to travel long distances in search of food. Unlike most Human-era bats, deathgleaners are only active in the daytime. They must wait until the sun has warmed the ground sufficiently to make use of warm air currents, or thermals, and soar over great distances.

When food is particularly scarce, deathgleaners go into a state of torpor, saving energy by slowing down their metabolism. This strategy for conserving energy ensured their survival when so many other animals and birds faced extinction. As the cold deserts spread over North America, the bats were able to fill the scavenging niche that had been vacated by buzzards and vultures.

Once the attack on the spink colony is over and the deathgleaners have eaten their fill, the leftovers are gathered up and carried off. The meat will be taken back to the roost where it will be shared with related members of the colony. This is not a new phenomenon. Human-era vampire bats also shared food, filling special reserves with the blood of prey, which could then be passed on to other bats that had not eaten. By sharing food in this way, the deathgleaners aid the survival not only of individuals but of the entire species.

left

Deathgleaners leave their roosts in the daytime, when they can soar over great distances on warm air currents rising up from the desert.

right

Human-era vampire bats fed on the blood of animals such as cows and horses. They shared food with other bats within the colony, forming tight blood-sharing bonds.

59

END OF AN ERA

THE ICE AGE IS COMING TO AN END. An increase in volcanic activity has released large amounts of carbon dioxide into the atmosphere, leading to a gradual warming of Earth. Even such a slight increase in global temperatures can cause the icecaps to melt and set off a far-reaching chain reaction. As the brilliant-white ice sheets retreat, revealing the dark earth and rock beneath, the surface of the planet becomes less reflective. This means that more solar energy is absorbed by Earth and less is reflected back into space. Thus, little by little, temperatures continue to rise.

The final glacial retreat happens relatively swiftly, over a period of 2,000 years. The icecaps contract, revealing the carved, scoured landscape beneath and leaving deep scars and heaps of displaced rubble over much of the northern hemisphere. The extremes of Earth's climate gradually soften, the blizzards cease and the polar icecaps melt back until there is little permanent ice left. It will be a long time before Earth sees another Ice age as severe as this one.

As Earth warms up, more and more water becomes available to life. The atmosphere can now hold more moisture, so humidity levels rise. The oceans are getting warmer, and as the sea water heats up, it expands, contributing to the already rising sea levels. Water spreads across continental shelves, bringing shallow seas to the edges of landmasses and, in some areas, reaching inland. Prevailing winds bring rain to warming continents. Plant growth returns to what were once arid regions. and tropical forests begin to flourish once more.

Many of the animals that had adapted to the bitterly cold conditions of the Ice age are unable to keep up with these rapid changes to their habitat and climate. The least adaptable of them die out: the large furry predators of the tundra and the highly specialized creatures of the arid deserts are particularly vulnerable. Only the most versatile species survive and evolve with the changing conditions. Gradually the world's climate stabilizes, heralding a benign period for Earth that will last for many millions of years. It will take time for life to recover from the climatic upheaval and mass extinctions of the past five million years, but eventually a whole new flora and fauna will populate a green and fertile land.

"This period coincides with an increase in volcanic activity. Volcanoes release a lot of carbon dioxide into the atmosphere, warming Earth and melting the ice. This sets off a cycle in which, as the ice disappears, Earth becomes less reflective. Consequently, less energy is reflected back into space and Earth gradually warms up, bringing the Ice age to an end."

Dr Roy Livermore
Paleogeographer
British Antarctic Survey

right

The end of the Ice age marks a new beginning for life on Earth. As the ice sheets retreat and climates begin to stabilize, the surviving animal and plant species will spread out and colonize newly-revealed habitats.

100
MILLION
YEARS

HOTHOUSE EARTH

"For millions of years since the last Ice age, Earth has enjoyed a warm, stable climate. Under such benign conditions, organisms can start to diversify and spread out, becoming extremely well adapted to particular niches where they fit in 'just so'. With no sudden environmental or climatic changes to spur sudden evolutionary change, the world becomes full of highly specialized organisms."

Professor Bruce Tiffney
Paleobotanist
University of California

THE WORLD MOVES ON. Along the vast underwater mountain chain that is the mid-ocean ridge system, molten lava rises up out of Earth's mantle and solidifies to form new ocean crust. Older crust is carried to either side of the ridge, causing the seafloor to spread and the continental plates to drift. By this continuous process of plate tectonics, the continents slowly move across Earth until they eventually collide with other continental plates.

As the continents meet and fuse together, volcanic mountain ranges are formed. The volcanoes spew carbon dioxide into the atmosphere, causing global temperatures to rise. The last Ice age ended five million years after the Human era. Now, 95 million years later, the world is a moist, warm place. Average temperatures are four or five degrees higher than they were during human times, and considerably higher than they were during the Ice age. There is very little permanent ice.

The melting icecaps have released large volumes of water into the oceans, producing sea levels more than 330 feet (100 meters) higher than during the Human era — higher than at almost any time in the planet's history. The Shallow Seas now cover any low-lying land. Only the highest uplands and mountain chains remain dry.

As Australia was carried north, it eventually collided with the eastern coast of Asia, throwing up a chain of mountains even larger than the Himalayas of the Human era. High up in these mountains lies the Great Plateau, a high-altitude habitat where the air is thin and the rocky slopes are populated by giant spiders and relic mammals. Antarctica has also moved north, its climate changing as it drifted into more temperate zones. The frozen wastes that once characterized this uninviting land have now been replaced by the warm, lush Antarctic Forest.

What was the Bay of Bengal is now a vast, brackish swamp, cut off from the sea to the south by a large chunk of Africa which has traveled across the Indian Ocean and fused to the southernmost tip of Asia. Covering hundreds of thousands of square miles, the Bengal Swamp is a hot, humid place where water is plentiful and vegetation abundant. It is a land of giant herbivores and strange aquatic predators.

Millions of years have passed since the last Ice age. During this time, the planet has luxuriated in a climate perfect for the development of life. The warm, moist conditions have spawned a great diversity of flora and fauna, leading to increasingly specialized behavior and adaptations. It is a hothouse world, brimming with life.

Changing globe

100 million years after the Human era, the icecaps have melted, causing sea levels to rise and covering much of the planet in shallow seas. Australia has collided with Asia, forcing up a huge mountain range between the two continents. Antarctica has moved north, its climate becoming warmer as it drifted into more temperate zones. Part of Africa has split from the rest of the continent and traveled eastwards across the ocean, eventually becoming fused to the southernmost tip of Asia.

Earth today
Earth is in a warm spell of an ongoing Ice age. A large amount of sea water is frozen into the icecaps that cover the North and South Poles.

in 100 million years
The icecaps have melted, sea levels have risen and global temperatures are several degrees higher than they were in the Human era. Earth is warm, moist, and luxurious, with a great diversity of species.

key
- land areas
- mountains
- upland
- forest
- swamp
- ocean
- shallow seas

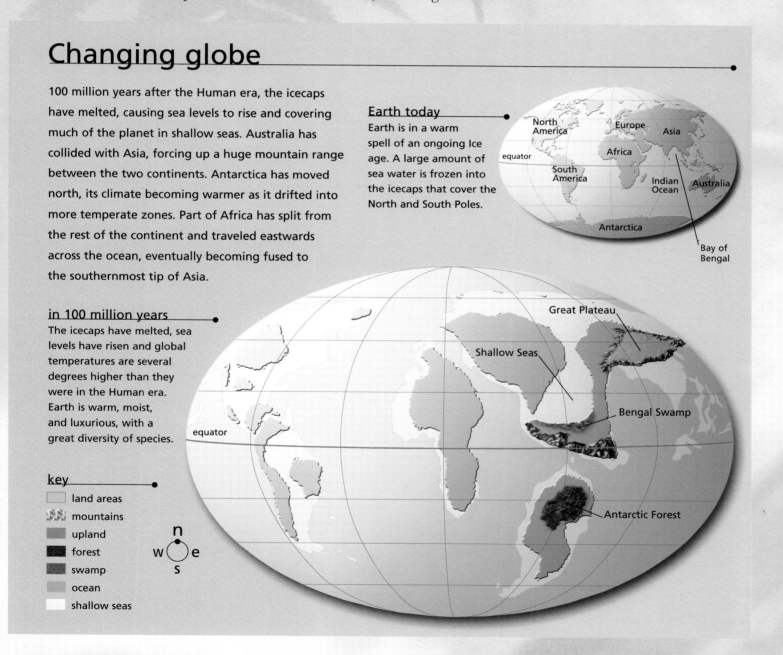

the SHALLOW SEAS

100 MILLION YEARS AFTER THE HUMAN ERA, a warm global climate has caused the polar icecaps to melt and sea levels to rise by around 330 feet (100 meters). Lower-lying parts of the continents are flooded and the oceans have spread southwards from the Arctic and eastwards from the Atlantic. Vast tracts of Russia are now almost entirely underwater. The Shallow Seas, which stretch across northern Europe and Asia, are punctuated by rocky islands – the peaks of mountains not yet covered by water.

The sun-filled, nutrient-rich waters of the Shallow Seas provide ideal conditions for the formation of reefs. Reefs are essentially calcium deposits, built up by generations of reef-building marine organisms. These organisms extract calcium dissolved in seawater and use it to lay down protective shells. Over successive generations, the shells and skeletons of the reef-building organisms accumulate to create a great edifice – a solid foundation upon which reef builders live and photosynthesize. This edifice is a reef.

The dominant reef-building organisms of the Human era were corals. Algae cells lived symbiotically with the coral, supplying oxygen and carbohydrates used in the production of calcium. With the last glacial period, five million years after the Human era, came massive climatic disruption. The seas filled with mud, depriving the algae of the sunlight they needed for survival. Without the algae and the essential nutrients they provided, the corals also became extinct. Now, tens of millions of years later, there are large areas of warm, shallow water and conditions are right for reefs to develop once more. This time, the reefs are formed not from coral, but from a prolific species of red algae.

Red algae were not always so prolific. Their spores can't swim, so the algae's fertilization technique was a matter of releasing millions of spores into the sea and relying on favorable currents to carry them to another plant. Now, 100 million years after the Human era, the red algae have evolved a sophisticated, symbiotic fertilization process by teaming up with a reef-dwelling animal. The algae offers up a protein meal and, as the animal moves from plant to plant feeding on this meal, it transports the algae's sticky spores. This new species of algae produces fewer spores and yet enjoys a much better fertilization rate. With such reproductive efficiency, the red algae have built up a successful reef system.

"100 million years in the future, warmer conditions will have melted the icecaps. Sea levels will have risen by up to 330 feet (100 meters) and low-lying land will be flooded. There will be huge expanses of shallow sea where sunlight will penetrate right down to the bottom. It will be a very vigorous and dynamic ecosystem."

Professor Paul Valdes
Paleoclimatologist
Reading University, UK

left

A young reef glider darts among the jagged peaks of the algal reef, feeding from the flower-like protuberances.

left

An adult reef glider, its gills
flowing behind in a colorful display,
paddles gently through the greenish,
sun-infused waters of the algal reef.

Shafts of sunlight slant through the clear water of the Shallow Seas, penetrating as far as the jagged spikes of the algal reefs which form the sea bed. Here and there, cup-like shapes sprout from the red surface of the reef. Swimming animals dart among the reefs and algal flowers.

Through the water sweeps a large form, silhouetted against the bright surface and throwing its shadow across the reef. From its bulbous, teardrop shape protrude three pairs of paddles. A long bunch of streamers trails behind. This is a reef glider, descended from small nudibranchs that were common in Human-era seas. At that time, they were unflatteringly referred to as sea slugs but, despite their ugly name, they were very dainty creatures, colorful and elegant, their gills forming fern-like arrays on their backs.

100 million years after the Human era, nudibranchs have evolved into substantial animals – adult reef gliders grow to the size of seals. Like their ancestors, they are brightly colored, their gills flowing behind them in a long train of silky fibers, like the tail of a bird of paradise. The reef glider's head sports a pattern of bumpy, scent-detecting chemical receptors, called rhinophores, and groups of eyes mounted on stalks. It has a horny, beak-like mouth. Unlike its ancestors, the reef glider no longer relies on simply crawling across the reef by expanding and contracting its body. It has developed paddles, which are fleshy extensions of its invertebrate body. Each pair beats in turn, propelling the animal slowly through the water.

To swim efficiently and to save energy, a large, slow-moving marine animal needs to be neutrally buoyant – able to float effortlessly in the water. Bony fish achieved this using their gas-filled swim bladder. Sharks kept their bodies lighter than sea water by means of low-density oils in their tissues. The reef glider's method is more sophisticated. By replacing the heavy sodium ions in its bodily fluids with lighter ammonium ions, the reef glider becomes lighter than the water around it. Staying buoyant requires large quantities of ammonium-rich fluid, hence the glider's bloated shape.

Adult reef gliders spend most of their lives in open water, coming to the reef to breed and to give birth. The algal reefs are a secure nursery for the baby reef gliders, which derive most of their nutrition from red algae. The juveniles are smaller, faster and more brightly colored versions of their parents. They flit about the surface of the reef, extending their long mouths into the cup-shaped structures formed by the reef algae.

In the base of the cups, the algae produce deposits of protein and carbohydrate. The baby reef gliders' beaked mouths reach into the cups to gather food. As they emerge, they are coated with sticky strands of reproductive cells. These are carried across to another cup and scraped off as the young animal feeds. Like Human-era bees, the baby reef gliders are acting as fertilizing agents for the algae.

This strategy of recruiting other marine life to help out in the reproductive process has evolved over millions of years. As evolution took its course in the fertile habitat of the Shallow Seas, algae split into two broad types. Those algae which released sperm cells into the open sea placed their survival at the mercy of the currents and fate. Meanwhile, algae that began to rely on unwitting feeding animals to distribute their sticky reproductive cells were soon able to breed more efficiently. Eventually, the sticky-sperm trait became predominant, and algae that did not have it died out. Evolution does not always follow the most straightforward route, but it always rewards reproductive efficiency, as it has in the case of red algae.

The reef-forming species of red algae went through a long period of evolutionary experimentation to perfect its reproductive strategy. Different kinds of swimming animals were used as fertilizing carriers, known as vectors, and the algae developed different forms of marine 'nectar' to act as bait for them. Finally, the flower-like structures were developed to tempt and feed the vector animals, and reproductive cells were produced within these. The female part of the 'flower' became more sophisticated too, evolving parts to remove the fertilizing cells from a visiting animal. Over time, juvenile reef gliders became the ultimate vector animal, and have been for millions of years.

Sea slugs

Reef gliders are descended from nudibranchs, known as sea slugs, which were common in the Human era. Nudibranch means 'naked gills' because sea slugs carried their gills as exposed tufts outside their body. Sea slugs were small, colorful mollusks, no more than an inch long. They were related to snails but lacked a protective shell. The bright colors of sea slugs served as a warning to potential predators that they were poisonous. They were able to isolate stinging cells from the sea anemones and sponges they fed on, and transfer the poison to their own skin. Thus, sea slugs transformed themselves into an unpalatable meal.

A tiny, exquisitely-colored sea slug crawls across the sea bed, its gills forming a tuft on its back.

Monumental reef structures, graced with pretty algal fronds and flower-like appendages, live in harmony with the elegant reef gliders, their vivid colors glinting through the green shallows. At times, the Shallow Seas resemble an idyllic underwater garden. But this is the wild, and in the wild there is always danger. As in most stable ecosystems, plants provide the initial food supply, having manufactured their own food using the energy of the sun to photosynthesize. Plant-eating animals exploit this food resource – often to the mutual advantage of both animals and plants. However, plant-eating animals are themselves part of the food chain, and in the warm waters of the Shallow Seas, they represent a tempting food supply.

As the young reef gliders cruise and gyrate among the red algae, something large and sinister glides slowly overhead. It casts a long shadow which creeps across the irregular surface of the reef. This is not the shadow of an adult reef glider, but something much slower, much bigger and much more dangerous.

The looming shape above has a peculiar translucent quality. It trails a forest of streamers and tentacles that brush across the surface of the reef, probing into crevices, questing around the red cups of the algae. The creeping darkness overhead throws the young reef gliders into confusion, but none seems aware of the real danger. Then one, more curious than the others, ventures too close to the bell-shaped end of one of the tentacles. In an instant, the tentacle tip opens like an umbrella, then snaps shut again. The baby reef glider is gone, sucked up into the body of a giant predator.

This floating menace is an ocean phantom. Its Human-era ancestors were siphonophores – a type of communal jellyfish, delicate and transparent in appearance but predatory in behavior. The most well-known of these was the Portuguese man-of-war. Each organism consisted of a collection of individuals – one provided a flotation chamber while others adapted as feeding organs, stinging organs or reproductive organs.

A floating colony

The Portuguese man-of-war was found in warm tropical seas during the Human era. It was a floating colony of individual organisms, called polyps, which each carried out a separate function – feeding, stinging and reproducing – while remaining dependent on one another for survival. A fourth type of polyp formed the crest of the man-of-war, which acted as a sail, changing shape to catch the prevailing wind and allowing the colony to navigate. Prey was captured in long, poisonous tentacles which could be regenerated by asexual budding, whereby the polyp would sprout a genetically identical extension of itself. The man-of-war fish fed off the poisonous tentacles but was immune to their toxins. In return for food, it acted as a lure to other fish.

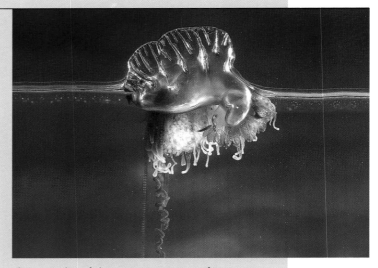

The tentacles of the Portuguese man-of-war contained one of the most powerful poisons known in Human-era marine animals.

left

The suction bell at the end of an ocean phantom's hunting tentacle closes around a juvenile reef glider.

overleaf

An ocean phantom in full sail glides menacingly over the reef, its grasping tentacles seeking out unfortunate victims from the sea bed.

Now, 100 million years on, the Portuguese man-of-war's descendants have grown to a huge size. A typical ocean phantom is over 30 feet (10 meters) long, 13 feet (four meters) wide and consists of many thousands of individuals. The largest part of it is the float. This looks like a giant mattress, made up of an assemblage of small air sacs. On its surface are a number of sails, turning and catching the wind to drive the animal along. These sails are gas-filled, but their walls contain a network of tubes which can be filled with water to control their shape. By filling different tubes, the ocean phantom can turn the sail to face any direction, catching the wind from whichever quarter it comes. When the water is withdrawn, the sails collapse.

The ocean phantom does not just drift downwind. When necessary it can tack like a yacht, moving its bulk against the wind. The force of the wind is counterbalanced by the pressure of the water against submerged 'keels'. There are also individuals in the colony that produce water jets, driving the phantom along when winds and currents are insufficient to do so. Other individuals act as rudders.

This sophisticated set-up carries the ocean phantom from one reef to another on its quest for food supplies. It can sense wind direction and bottom depth, and so avoids dangerous shallows, beaches and exposed rocks where it might become trapped or damaged. When not feeding or avoiding hazards, the phantom simply drifts along with its sails collapsed and its underwater appendages retracted. It traverses the seas like an enormous piece of lifeless flotsam. Only when it is hungry, or when it drifts towards the reefs, does the whole colony spring into action.

A complex sensory array is also used to detect feeding grounds and potential danger. This system can analyze wind strength and the position of the sun, information which allows the animal to navigate. Human-era siphonophores possessed a simple neural network, connecting individuals of the colony to one another. The sheer size of the ocean phantom and the degree of communication required for so many individuals to function in harmony mean that, in many ways, this mass of living matter resembles a giant, floating brain.

71

Above the water line, the ocean phantom's exposed surface is covered by algae. Like the reef algae, these organisms build structures from calcium, forming small trunks upon which the algal strands cling, streaming in the wind. The algae are a kind of farm, providing much of the nutrition for the siphonophore colony. Carbohydrates generated by the algae through photosynthesis are carried throughout the colony by a vascular system which takes food to every individual member. In return, the algae are provided with a safe base and even a supply of fresh water. When rainfall does not provide this, specialist individuals in the ocean phantom can desalinate seawater and pump it to the upper surface.

Algae-generated sugars and starches do not provide all the nutrition the colony needs. For protein it must hunt animals in the waters beneath. This is why the ocean phantom presents such a threat to the reef-living creatures. When the dark shape drifts overhead, all except the youngest and most inexperienced of the Shallow Sea swimmers make themselves scarce.

The ocean phantom uses suction bells to hunt. At the end of each hunting tentacle there is a spherical structure with a downward-pointing mouth. In the tentacles above, muscle fibers operate like bellows, expanding and contracting to open the bells. Small water jets are used for more accurate, localized movement. Around the mouth of each bell, there is a ring of stalked eyes and feathery sensors. Such a structure allows each tentacle to function and hunt independently.

This apparatus is dragged across the surface of the reef until it detects prey of the right size. The bell then opens and closes, its pulsing action driving the whole unit into an attack position. With a sudden dilation, the prey is drawn into the bell and trapped, to be digested at the ocean phantom's leisure. The nutrients extracted are then pumped up the tentacle and distributed to the whole colony through its vascular system. Nitrates from the prey are also delivered to the algal meadow on the exposed surface of the ocean phantom, providing additional nutrition for the plants.

left

The ocean phantom conceals a secret weapon inside specially adapted tentacles. In return for food and shelter, an army of vicious sea spiders, called spindletroopers, is on hand to defend the phantom from predators.

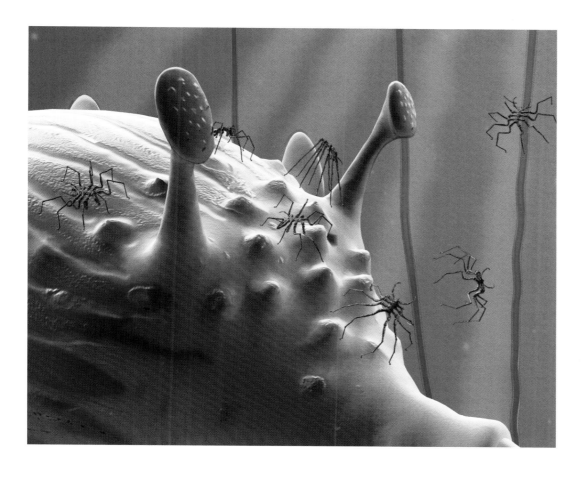

left

Spindletroopers defend an ocean phantom from a marauding gang of reef gliders. The sea spiders' sharp fangs and slashing claws swiftly repel the hungry predators.

Underpinning the ocean phantom's sophistication and the astounding specialization of its constituent parts is an inherent flexibility. The whole structure of the creature is modular. It can exist as a huge mass or as smaller units, provided each unit has the full range of individuals it needs to survive. After a heavy storm, a large ocean phantom may have broken into many smaller parts, each of which will eventually regenerate. The ocean phantom is not invulnerable, though. Drifting along in the open water, even this huge creature is open to attack.

Adult reef gliders which, as juveniles, were prey to the ocean phantom, are now its chief predator. As they move into adulthood, reef gliders must leave the reef in search of a larger source of protein. The ocean phantom is at the top of the menu. Circling their victim like slow-moving sharks, the reef gliders home in on the floating mass. Their horny beaks tear into the air sacs, the tentacles, the rudders

and the keels – in fact, every part of the ocean phantom beneath the water line is a potential meal. Chunks bitten from the ocean phantom are the adult gliders' main food supply. In the face of such an assault, the phantom has evolved a brilliant symbiotic defense mechanism.

Some of the suction bells have ceased to function as hunters. Instead, they are troop carriers. Should a reef glider be unlucky enough to brush against one of these modified bells, a horde of spindly legs, slashing claws and slicing fangs emerges. It is an army of spindletroopers, come to the rescue of its host.

The spindletrooper is a species of pycnogonid, commonly known as a sea spider. During the Human era, shallow-water sea spiders were small, rarely exceeding a few inches. With a leg span of 12 inches (30 centimeters), the spindletrooper is far larger than most of its ancestors.

75

Before forming their symbiotic partnership with the ocean phantom, spindletroopers lived on the algal reefs, occasionally raiding an ocean phantom's suction bells for food. Soon they began to remain safely inside the bells for longer periods of time. Eventually, they adapted to fold up and fit neatly inside a suction bell, which in turn modified itself to house and feed the spindletrooper.

When hungry, a spindletrooper scratches at the walls of its home bell, stimulating it to regurgitate food from the rest of the colony. In return, the spindletrooper provides defense, coming out to fight when the colony is threatened. Spindletroopers have large jaws full of sharp fangs, capable of delivering an injection of painful poison. With these and their long claws, they slash at anything that attacks their ocean phantom home.

The Shallow Seas and their colorful, complex reefs have presented a stable environment for a long time. They have persisted for so many millions of years that effective living systems, established early on in the history of the habitat, have been able to survive without serious challenges. As a result, creatures' body shapes, feeding methods and relationships have modified only slightly over time. However, these warm seas have allowed one thing to change a lot, and that is a creature's size.

Over time, certain evolutionary trends have become apparent. One such trend has occurred again and again in response to attack-and-defense situations: each party gains the advantage over the other by evolving to become larger. Bigger predators evolve to overcome the defenses of their prey. In turn, prey also becomes larger, making it more difficult for predators to attack. This explains why the Shallow Seas, though similar in many ways to the coral reefs of the Human era, are full of huge animals. Creatures that were small 100 million years ago, such as sea slugs, siphonophores and sea spiders, have evolved into giants.

More fascinating still is the development of symbiotic associations between species. The ocean phantom has developed two such relationships. Above the water level, it provides secure real estate for the algal meadows which flourish on its back, and in return receives a constant food supply. Below water, a defending army of spindletroopers lives inside the bell-shaped tips of its tentacles. In return for the provision of nourishment and housing, the spindletroopers protect their host from the potentially devastating attacks of adult reef gliders. These unlikely pairings have proved to be highly effective partnerships: not only are they borne out of mutual need, but they are equally beneficial to both parties.

right
In the aftermath of a storm that has pounded the Shallow Seas and their rocky coasts, the broken remains of an ocean phantom float away in the sunset. Each section will regenerate and begin its own quest for survival.

the BENGAL SWAMP

THE CONTINENTS HAVE TRAVELED VAST DISTANCES since the Human era. A massive raft of land which broke away from Africa along the Great Rift Valley has moved across what was the Indian Ocean and collided with the southernmost corner of South East Asia. As the two landmasses came together, a vast inland sea was created between them in the area that was once the Bay of Bengal. The massive forces created by the colliding tectonic plates buckled the landmass, giving rise to a volcanic mountain range along the line of fusion. Over time, the inland sea became almost entirely cut off from the oceans to the south.

Water run-off from the mountains formed rivers which washed fertile sediment into the landlocked sea. Eroded material from the newly-exposed rocks was carried downwards, filling the basin and making it shallower and rich in nutrients. Gradually, the inland sea diminished, fresh water from the mountains mixed with the residual salt water, and a vast, brackish swamp was formed.

100 million years after humankind, the Bengal Swamp covers hundreds of thousands of square miles. Sediment carried by slow-moving channels and meandering rivers makes the water thick and impenetrable to light. Sedimentary deposits have formed a series of ox-bow lakes and backwaters separated by muddy islands and flats. The Bengal Swamp is comparable in appearance to the great lowland coal swamps of the Carboniferous period, 300 million years before humans.

The climate of the Bengal Swamp is hot. Its proximity to the equator and the shelter provided by the surrounding mountains mean that average temperatures are about over 100°F (40°C). Water is plentiful, running down from the mountains in an intricate network of rivers. Humidity is extremely high, averaging 99 per cent all year round. The muds and soils are constantly replenished by nutrient-rich volcanic ash.

A greenhouse environment like this is an ideal place for vegetation to grow. Plant life chokes the waterways and spreads across the lakes. Thickets of tropical plants clothe the sandbanks and deltas. Tightly-spaced trees stand where any land is solid enough to hold them, spreading deep canopies of branches and leaves overhead and stabilizing the mud with their network of roots.

"100 million years in the future, one of the environments we're looking at is a huge, near coastal swamp. It is partially brackish as influence from the ocean, but a huge amount of fresh water also flows into it. Earth at this time is like a vast greenhouse, and so this swamp is high in moisture and very rich in life."

Professor Bruce Tiffney
Paleobotanist
University of California

left

Humid, hot and rich in nutrients, the environment of the Bengal Swamp supports an abundance of plant and animal life.

left

A lurkfish lies in wait, its amazing camouflage making it almost invisible. When it finally pounces, the lurkfish subdues its prey with a huge electrical charge.

A host of dangerous creatures dwell in the murky backwaters and shallows of the Bengal Swamp, beneath the tangle of thick, choking vegetation. Perhaps the most dangerous of these is the lurkfish.

At first glance, the scaly surface of the lurkfish could be mistaken for the lumpy bark of a decaying tree trunk, its fins and spines for broken fronds and branches. At 13 feet (4 meters) long it certainly resembles a large log. However, this is no piece of wood, but a very sophisticated and powerful hunter.

For days the lurksfish lies in the water, not feeding, not moving. Being cold-blooded, it only needs to eat occasionally. It is an ambush predator, lying in wait. The mouth of the lurkfish splits the whole of its broad snout and from its thick lips hang short barbels that are sensitive to any movement in the water. More barbels sprout from

its face. Tall and branched, they mimic the thick tangle of water weeds covering the swamp. But the barbels are not just a means of camouflage. Like the electric catfish which populated the Congo and Nile rivers during the Human era, the lurkfish is able to generate a powerful electric field. The barbels form an electrical sensory net, capable of detecting even the smallest movement of potential prey in the water nearby.

After days of pretending to be a tree trunk, the lurkfish is preparing to attack. It holds back, waiting until an unsuspecting victim moves to within striking distance. At the sides of its body, the lurkfish's pectoral fins carry muscular spines that dig into the underlying mud and slowly raise the massive front end from the swamp bed. Then its tail paddle, composed of a caudal fin, rear dorsal fin and anal fin, sweeps suddenly and powerfully and thrusts the creature's hungry bulk forwards.

There is a spray of mud and water. The lurkfish's mouth opens, engulfing its helpless victim. The lurkfish then descends into the thick, muddy water of the swamp. Waves caused by the sudden attack wash against the trunks of surrounding swamp trees. Ripples spread across the opaque water and die away. Everything is as it was once more.

This was a simple hunt for the lurkfish. Small prey is hardly a challenge for this swamp monster. Other means must be employed to subdue larger prey capable of putting up a strenuous and sustained fight. While a larger victim might not escape, it could certainly do severe damage to the lurkfish before succumbing. Like a Human-era crocodile, the lurkfish has a mail of thick, protective plates covering its body beneath its skin. This armor shields the lurkfish's internal organs, but its external fins, barbels and sensors are vulnerable in a fight. The lurkfish has another weapon at its disposal. Its electrical system is not just a sensory device but is also used in hunting. A lurkfish can overcome large and combative prey by administering a powerful electric shock to its victim.

Along each side of the lurkfish are stacks of electrical muscle blocks called electrocytes. Separately, these electrocytes are able to generate only a small charge. But hundreds in series along the fish's 13-foot (four meter) long body can generate a huge cumulative charge of over 1,000 volts – double that of a Human-era electric fish. Whatever the size of the prey, it will be paralyzed by an electric shock of such power. Great jaws will close around the motionless victim, rows of fine teeth will sink into its flesh, and the lurkfish will disappear to dine at its leisure.

Once the lurkfish has struck, the surrounding area of swamp seems to come to life. The disturbance in the water, the splashing noises and electrical signals, all send out the message that danger has passed. The lurkfish has hunted successfully and will rest while it digests its prey.

The Bengal Swamp has become so dangerous, so full of large predators like lurkfish, that some creatures have taken refuge out of the water. Indeed, the atmosphere of this hot swamp is so humid that many aquatic animals can spend time on land and not suffer any discomfort.

Electric fish

Electric fish generate voltage in small muscle blocks, called electrocytes, which lie in rows along the length of the body. Each electrocyte generates a small electrical potential. Like batteries in a series, the muscle blocks build up a cumulative charge. The larger the animal, and the longer the series of electrocytes, the greater the electrical potential. A Human-era electric eel measuring over three feet (one meter) in length could generate up to 600 volts of electricity. At 13 feet (four meters), the lurkfish can easily double that, delivering an electric charge of over 1,000 volts, which it uses to paralyze its prey.

Human-era electric eels used low-intensity electrical impulses for navigation and hunting. High-intensity impulses were used to stun or kill prey.

One creature to have taken advantage of the relative safety of land is the swampus. It is a cephalopod, a distant cousin of the marine octopuses that were so prevalent in the Human era. The swampus has evolved to survive on land, but only for limited periods of time. It is unable to breathe properly out of water and relies on finite stores of oxygen in its tissues and blood. Once these reserves have been depleted, it must resubmerge itself in the swamp water to replenish its oxygen supply.

A swampus emerges from the ferns and tangled creepers of a thicket and slithers into the water. From over the moss-slicked surface of a fallen tree comes another. The water is safe for a while and its inhabitants are returning. Swampuses that have used up the reserves of oxygen stored in their tissues and blood gradually become more and more desperate to submerge and breathe properly. For a few moments, the banks and shorelines are alive with movement as large numbers of swampuses come out of hiding and plunge into the still waters of the Bengal Swamp.

In most respects, the swampus looks like the Human-era octopus from which it has evolved. One difference is immediately apparent: at first glance this creature appears to have four arms instead of eight. This is because four of its original arms have evolved into weight-bearing pads which it uses to move over land. They function like the foot of a snail, carrying the animal along by a rippling action of the muscles. The suckers of the arms have developed into horny ridges that grip the ground beneath them. The long arms at the front of the swampus are also used in locomotion, reaching out to grip logs and trunks, and then pulling the creature along. It is a clumsy movement, but sufficient to carry the animal's 44 pound (20 kilogram) weight through the thick vegetation of the humid swamp.

The swampus can survive on land for up to four days at a time, safe from the electrical attacks of lurkfish. However, it must return to the water to replenish its oxygen supply and carry out other vital functions, such as mating. The murky water is no place to rear a family, though. Once

Adapting to life on land

The swampus has successfully adapted to life both in and out of the water. When it goes ashore to escape predators or tend to its young, it survives by means of oxygen stored within its body. Once these oxygen reserves are depleted, the swampus must return to the water to breathe properly once again.

Mudskippers, a type of fish found in the mangrove swamps and tropical shorelines of the Human era, were also able to survive on land. Underwater, mudskippers breathed through gills like other fish. On land, they derived oxygen from water stored in enlarged gill chambers which locked shut, keeping their gill filaments moist and oxygenated.

Human-era mudskippers were able to survive for several days out of the water.

left

The swampus uses its forearms to grab items of vegetation and pull itself along through the swamp.

she has mated, the mother swampus clambers out of the water and into the surrounding vegetation to find a safe place in which to deposit her fertilized eggs.

The mother heads for a spray of lily plants which serve as a nest in which the eggs can hatch. This nursery plant produces a flush of large, vase-shaped leaves close to the ground. Rainwater partially fills newly unfurled leaves, creating freshwater pools. The leafy pools act as protective cradles for the eggs and young of the swampus.

A pregnant female swampus returns to the same patch of lily plants year after year. She deposits her eggs in the rainwater pool and stands guard over them while they hatch, protecting both flower and brood. Throughout evolutionary history, cephalopods – octopuses, squids and cuttlefish – have been unable to cope with fresh water, and the young swampuses are no different. To solve this problem, the female has evolved a means of changing the chemical composition of the pool. From time to time,

she urinates in the water in order to recreate the salty environment of the swamp. Then, by splashing about in the water with her arms, she oxygenates it and her offspring are able to breathe.

The mother must also breathe and, to do so, she must leave her offspring and return to the swamp. Her young are not abandoned, but left in the charge of other females who have deposited their eggs in nearby plants. They continue to monitor the chemistry and oxygen content of her nursery as well as their own.

On her return, the mother recognizes her baby-sitter by means of a complex communication system. Her arms are sensitive to touch and chemical signals, and so she is able to recognize the particular taste and texture of other individuals. This system works well because related females bring up their families at the same time in other lily plants of the same patch. Swampuses are highly social creatures.

Like its ancestors – the octopuses of the Human era – the swampus is a master of camouflage. By stimulating sacs of pigment in its skin, called chromatophores, it can change its color to blend in with its surroundings. The knobbly surface and serrated edges of its arms mean that it can lose itself in the tangle of vegetation and detritus that litters the swamp floor. In this way, it can become almost invisible. Of course, this ability to mimic its surroundings is not just used for self-defense. The swampus is an accomplished predator and uses its flexible, prehensile arms to catch large insects and small vertebrates which inhabit the undergrowth of the Bengal Swamp.

The swampus has one more weapon in its considerable arsenal. Many Human-era octopuses had a venomous bite which they used to subdue prey. 100 million years after humankind, the swampus employs a similar strategy, and the story of how it obtains the toxins for its noxious weapon is one of remarkable co-evolution.

The lily plant's vase-like basin contains a growth of bacteria. As the baby swampus grows up in its organic nursery, it gradually ingests the bacteria. Once ingested, the bacteria form the basis for the swampus to generate its own venom. Babies and adults use the venom to defend their nursery from large herbivores which feed on the plant. In return for a safe home in which to grow up, the swampus provides protection for the plant. Any animal grazing near the swamp lily is swiftly repelled by a nasty injection of poison.

right

A family of swampuses gathers around a lily plant. The females lay their eggs and nurse their young in the vase-like basin of the plant.

In an environment such as the Bengal Swamp, where vegetation flourishes, the presence of large herbivores is no surprise. The largest of these is the toraton.

The toraton is the latest member of the ancient tortoise family. Since the Triassic period, tortoises and turtles have thrived. Their basic shape and lifestyle were so successful from an early stage in their history that they hardly evolved. Now, 100 million years after the Human era, conditions have encouraged a huge increase in their size, enabling them to exploit the incredible mass of vegetation that now exists in this greenhouse environment.

The toraton is a cold-blooded reptile, so it does not face the problem of overheating that would confront a similarly sized mammal. With a body weight of up to 120 tons, this is the biggest animal ever to have walked Earth, bigger even than the greatest of the dinosaurs. Due to its sheer size, the toraton has no predators and no longer has any need for a shell. Small sections of the ancestral shell do still remain, however, forming an external support for the creature's muscles, which its feeble ribs and vertebrae alone cannot carry.

The toraton eats constantly. A body this huge requires a large intake of food and the toraton consumes about 1,300 pounds (600 kilograms) of plant matter each day. Huge jaw muscles support a scissor-like beak which rips vegetation from the trees. The toraton does not chew, but grinds up its food in a muscular stomach, or gizzard. The rear part of the digestive system is a gut where bacteria break down remaining plant matter. This digestive system allows the toraton to eat virtually any kind of vegetation.

Young toratons emerge from eggs so tough that the mother helps them out by cracking the shell with her beak. Youngsters are cared for by their parents for the first five years of their lives. With no predators to threaten them, healthy toratons can live to the ripe old age of 120 years. They are the true colossi of the Bengal Swamp.

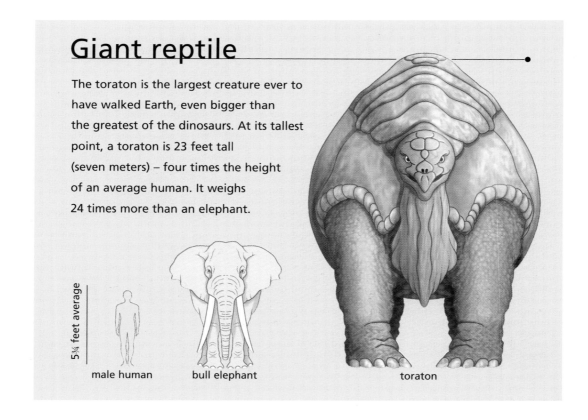

Giant reptile

The toraton is the largest creature ever to have walked Earth, even bigger than the greatest of the dinosaurs. At its tallest point, a toraton is 23 feet tall (seven meters) – four times the height of an average human. It weighs 24 times more than an elephant.

5¾ feet average

male human   bull elephant   toraton

right

Toratons walk in a slow and lumbering fashion, eating vegetation wherever it grows.

86

the ANTARCTIC FOREST

ANTARCTICA NO LONGER LIES OVER THE SOUTH POLE. The tectonic plate on which it sits has been moving northwards for 100 million years, gradually creeping towards the southern edge of Asia. This plate has carried the continent out of the polar zone, through the temperate zone and across the southern desert belt. Hard as it is to imagine, Antarctica now lies partially in the tropics.

The Antarctic icecap melted as soon as the landmass began to move into warmer latitudes. The continent's northern portion now lies well within the tropical zone of Earth's climate, where the converging trade winds bring warm rain all year round and the sun shines directly overhead. These are the ideal conditions for plant growth, and this is where the forests of the new Antarctica are located.

As a frozen continent, Antarctica was an inhospitable environment for life. In the Human era, it was home to few indigenous species. Plant life consisted almost entirely of lichens, mosses and algae. There were no land-living vertebrates and the few species of sea birds which inhabited the coastlines of the continent were the most successful animals. As it moved into more temperate regions, however, Antarctica became a much better prospect as a home.

Plants were the first of the newcomers: winds brought seeds and spores from South America in the east. Those seeds which survived the journey gave rise to adaptive radiation, where a variety of species evolved from a single ancestral species, each specifically adapted to the conditions they found on the Antarctic continent.

Spiders and insects were the next to discover the new habitat. Being lightweight they were carried on the winds, just as the plant seeds were. They settled easily among the newly evolving plant life. The first vertebrate settlers were birds, their powers of flight enabling them to cross the oceans to reach the isolated continent. And they brought with them yet more seeds and insects.

These new species of birds joined the descendants of the sea birds which had populated frozen Antarctica since the Human era. Having inhabited the old ice continent, the sea birds were ideally positioned to exploit the changes in their environment. This new, temperate land was a paradise by comparison and offered numerous and diverse possibilities.

"100 million years after the Human era, Antarctica is sitting very much in the tropics. It's obviously going to be hugely different to what we think of now. Surrounded by warm oceans, conditions will be very wet, ideal for a tropical rainforest. You can think of Antarctica as the new Amazon."

Professor Paul Valdes
Paleoclimatologist
Reading University, UK

left
The northern part of Antarctica is covered in lush tropical rainforest, which is home to many new species of plants, insects and birds.

left

The roachcutter uses its keen eyesight to spot prey. Its wings are short and broad, giving it great maneuverability in the dense forest.

One group of Antarctic sea birds that fared particularly well in the new conditions were petrels, such as shearwaters, fulmars, and albatrosses. In the Human era, petrels bred mostly on islands in the southern hemisphere, although some breeding grounds were as far north as the Caribbean. They also wandered widely at sea during the non-breeding season. Already capable of adapting to different climates, petrels remained and diversified as Antarctica gradually moved north, evolving to suit the changing conditions. They became the most varied and widespread group on the Antarctic continent, radiating to fill new evolutionary niches and becoming increasingly difficult to dislodge.

Now, 100 million years after humankind, the Antarctic continent boasts many species of bird. There are birds with long, narrow wings that soar over the land, small birds with short, broad wings that can maneuver easily in confined forests, and even flightless, ground-living birds. The majority of these are descended from petrels.

The most widespread group of petrel descendants are flutterbirds. These small forest-dwellers abound in the tropical rainforest of the northern part of the continent. The roachcutter is typical of flutterbirds. About the size of a Human-era sparrow, the roachcutter's wings have a high aspect ratio, meaning that they are short and broad, perfect for making tight turns. Feather tips are splayed out like fingers to manipulate the passage of air and increase maneuverability. The bird's eyes make it easy to distinguish from other flutterbirds – they are mounted on turrets.

With its small size and short wings, the roachcutter can hover like a bluetit, scanning the tree trunks for insects with its turreted eyes. Its beak is extremely tough, able to crush the hard outer cuticles of the insects on which it feeds. Even though it is adapted for slow navigation of tight spaces, the roachcutter is capable of reaching high speeds. At the sound of an approaching predator, it adjusts the angle of its wings and darts away between the trees. Speed and agility are good methods of defense, but another species of flutterbird has developed a truly impressive defensive strategy.

A bird slightly bigger than a roachcutter, with bright orange flashes on its wings, hovers in front of a forest tree. It dips its head repeatedly into the flower from which it appears to be feeding. Suddenly, there is a hum and flutter as the residents of the forest canopy flee an approaching predator. But rather than darting away to safety, this bird faces the danger. As the predator draws near, the bird lowers its head. Then, at the last possible moment, it sprays a hot, corrosive acid from its nostrils. This is the spitfire bird.

The spitfire bird was not feeding from the tree flower, but gathering a chemical from it. Like its cousin the roachcutter, the spitfire bird actually feeds on insects. The flower it was particularly interested in was that of the spitfire tree. The male and female of this type of tree produce different chemicals. Both are harvested by the spitfire bird and stored in a compartment of its throat, known as the crop. When faced with danger, the bird releases the chemicals, mixing them in a chamber in its nasal cavity and adding an enzyme from a gland in its skull. The enzyme unbinds the chemicals, which react violently with each other to produce the corrosive acid. The spitfire tree also benefits from the spitfire bird's plundering of its resources. As it hovers between the male and female trees, dipping its beak into their flowers, the bird aids pollination.

Another member of the flutterbird family, the false spitfire bird, uses an altogether more passive but no less effective form of defense. In appearance, it is almost identical to the spitfire bird. Unlike the spitfire, however, the false spitfire bird is harmless.

Chemical defense

The spitfire bird is not the only animal to use chemical weapons as a defense from predators. In the Human era, bombardier beetles employed a similar strategy, spraying hot, toxic chemicals at attackers with deadly accuracy. Chemicals secreted by the beetle collected in a reservoir within its abdomen. The reservoir opened through a muscle-controlled valve to a reaction chamber, where a series of reactions activated the chemicals, bringing the mixture to boiling point. The build up of gases released by the vaporized fluids forced the valve closed, sending the boiling-hot fluid out of the beetle through openings at the tip of its abdomen with a loud popping sound.

A Human-era bombardier beetle sprays a hot corrosive liquid at attackers.

The false spitfire bird avoids danger by mimicking the appearance of its more dangerous cousin, the spitfire bird, perfecting its disguise down to the orange flashes on its wings. This phenomenon is known as Batesian mimicry, where one species imitates the appearance of another to benefit from its attributes without actually possessing the attributes itself. But what predator do the flutterbirds fear?

Imagine a large, streamlined and voracious wasp. Compound eyes give it 180 degree vision and vicious jaws make it capable of devouring the toughest prey. Pairs of legs at the front and rear of the insect are equipped with strong gripping claws. A middle pair of legs come together to form a harpoon, barbed at the end. Now imagine that this thing is as big as a kestrel! It is a falconfly – the dreaded enemy of flutterbirds.

Insects are limited in their size by two factors: the mass of muscle needed to support a solid outer cuticle, or exoskeleton, and the ability of the creature to get oxygen into its tissues without lungs.

The atmosphere of the world 100 million years after humans is richer in oxygen than it was during the Human era. This higher concentration of oxygen makes a larger insect body possible.

During the Carboniferous period, 300 million years before humans, the atmosphere was similarly rich in oxygen and giant land arthropods flourished. The Carboniferous coal forests boasted dragonflies as big as magpies, scorpions the size of rats, and millipedes as long as vipers. In the Antarctic Forest, 100 million years after humans, wasps as big as birds of prey make light work of killing flutterbirds.

The falconfly cruises through the forest. It sees an unwary flutterbird hovering and feeding, and dives to the attack. The hooked legs seize the bird, grasping its body through its feathers, and the second pair of legs lance deep into its internal organs. With a squawk and a flurry of feathers, the two tumble through the branches and undergrowth and crash to the ground. There, the falconfly rips the flutterbird to pieces.

left

The falconfly is a type of wasp the size of a Human-era kestrel. Its grasping claws and barbed lance make it a formidable hunter.

left

A spitfire bird hovers about a spitfire tree, unaware of the approaching danger. Behind it is a falconfly, zooming in for an attack.

The giant insect has a family to feed. The falconfly has three or four burrows, each one containing a single developing larva. In the Human era, insects had several strategies for reproduction: some laid vast numbers of eggs, of which only a few survived; while others laid very few eggs but took special measures to ensure a high rate of survival. The falconfly pursues the latter course. It knows exactly where each of its larvae is hidden. It butchers the flutterbird and shares out the lumps of meat among its family.

The falconfly is not the only insect enemy of flutterbirds. The high oxygen concentration of the atmosphere has meant that insects are taking over from vertebrates as the dominant land-living animals. There are others, large and deadly, who are making the most of their time at the top of the food chain.

The spitfire beetle is one such insect inheritor of Earth. Apart from their bright red and yellow coloring, spitfire beetles resemble a typical Human-era beetle, with a head, thorax, diaphanous underwings and elytra (hard front wings). Individually, they are unremarkable. Collectively, they are capable of an ingenious form of ambush hunting.

The carnivorous spitfire beetles spend most of their lives in groups of four. They position themselves on the trunk of a spitfire tree, standing head to head in a cross formation. Spreading their wings, they are suddenly almost indistinguishable from the flowers of the tree itself. In this formation, the heads and thoraxes look like the flower's center, the antennae resemble the stamens, and the brightly-colored elytra the petals.

Motionless, the spitfire beetles wait, mimicking the flowers of the spitfire tree. Their intended prey is a spitfire bird, which is hovering about the tree and collecting chemicals. As it moves in, the beetles leap into action, seizing the bird before it can bring its defenses into play. Grasshopper-like hind legs propel the attack and strong jaws and grappling-hooked claws on the forelegs crunch into the bird's body. The carcass of the dead spitfire bird is then eaten by all four spitfire beetles.

While such co-operative mimicry was not common in the Human era, it was known. The larvae of the tortoise beetle exhibited similar behavior. These disc-shaped grubs would remain in clusters after hatching and react synchronously to anything which approached them by moving the tips of their tails upwards and mimicking the shape of a large spider. As the larvae's only predators were spiders, this proved an extremely effective form of defense and became an evolutionary success.

At the end of the spitfire tree's flowering season, when there is nothing more to tempt the spitfire birds, the colonies of beetles disband and disperse, looking for mates. After mating, the male beetle dies. The pregnant female flies around among the spitfire trees, laying clutches of four eggs beneath their bark. Having laid all her eggs, she too dies. The following spring, when the eggs hatch into the four individuals that form the flower imitators, the spitfire beetles will once again lie in ambush for a roving spitfire bird.

In the Antarctic Forest, the insects' time has come. Insects rival vertebrates in size and sometimes even surpass them. Whole new living strategies based on this reversal have evolved. Insects are brash and arrogant hunters, while birds have become small and furtive. Flutterbirds may still prey on insects, but the forest canopy is full of powerful, predatory insects evolved to prey on them.

Collective mimicry

Nature is full of imitation, but highly intricate and co-operative mimicry such as that practised by the spitfire beetles can only come about when conditions have remained stable for a long period of time.

The carnivorous larvae of the Human-era blister beetle employed a strategy of co-operative mimicry. A large number of larvae would form themselves into the shape of a female bee. Male bees, confusing the larvae for a female, then attempted to mate. The larvae would cling to the male bee's chest and be carried to a real female, where they could feed on her eggs.

Larvae of the blister beetle mimic the appearance of a female bee.

right

A unit of four spitfire beetles mimic the flower of the spitfire tree to attract unwary spitfire birds.

the GREAT PLATEAU

SINCE THE LATE EOCENE EPOCH, about 45 million years before the Human era, Australia became separated from Antarctica slowly north across the Pacific Ocean towards Asia. Where the two continental plates met, one was pushed below the other, creating a subduction zone to the southeast of the Asian landmass. As the ocean lithosphere – the rigid outer layer of Earth – was drawn down into the mantle and melted, new magma was produced, resulting in large amounts of volcanic activity.

Now, 100 million years after the Human era, Australia's short life as a single continent is over, and it has finally fused with the southeastern edge of Asia. Seafloor sediments and rock between the two landmasses have been compressed, sheared, ground together and thrust up into a massive mountain chain. This new chain exceeds the proportions of the Himalayas, the highest mountain range of the Human era.

Like the Himalayas in their time, these new mountains continue to rise. As the tectonic plates crush against one another, they simultaneously compress the rock downwards into Earth's mantle and upwards into the sky. Further compression has raised a large block of South East Asia to form the Great Plateau, the broadest tract of uplands on the surface of the planet. This immense plateau, surrounded by mountains, towers over the shallow shelf seas which cover much of the landmass.

Newly-formed mountains are sharp and jagged. It takes time for the constant assault of rain, wind, frost and running water to erode them into rounded shapes. 100 million years after humankind, the Himalayas are mere hills – undulations in the center of the continent. The Great Plateau, on the other hand, consists of ranges of pointed pinnacles and knife-edged crests dropping away into slopes of fragmented rock and scree. The valleys and basins between the ridges have filled with newly-eroded debris and formed upland plains, surrounded by peaks reaching up to 33,000 feet (10,000 meters) – higher than any mountains of the Human era.

How will life survive at this altitude? The climate of the weather-beaten peaks of the Great Plateau will certainly be harsh, but Earth 100 million years after humans is warm and volcanic activity has thrown large amounts of carbon dioxide into the atmosphere, making survival easier. There are ample resources for life to flourish.

"100 million years in the future, the climate of this high mountain plateau will be extreme. But, because the future Earth will be warmer, and carbon dioxide levels higher than in the Human era, conditions won't be anything like as harsh as, say, the high Tibetan plateaus of the Himalayas. It will be more possible for life to exist in these high mountain ranges."

Professor Paul Valdes
Paleoclimatologist
Reading University, UK

left

The Great Plateau rises up out of the sea, towering 33,000 feet (10,000 meters) above sea level. These inhospitable slopes, full of unstable rocky debris, are home to a number of species.

97

The Great Plateau, this system of high plains and basins, hemmed in by the highest mountains in the world, is not the dry, cold desert one might expect. Back in the Human era, high-altitude mountain systems such as the Himalayas were home to little more than hardy desert herbs, shrubs and small rodents. Not so the valleys and plains of the Great Plateau, 100 million years on. These are rolling grasslands.

At the outer edges of the Great Plateau, the steep, debris-covered slopes are swept by winds bringing seasonal rains up from the Shallow Seas. The heavy rainfall and loose soil make for an unstable surface, prone to mudslides and rock falls. However, in many areas the surface is stabilized by plant life evolved to cope with just such conditions.

This is the age of grass trees. Grasses are hardy plants which reproduce sexually by dispersing their seeds, and asexually (or vegetatively) by spreading out a network of underground stems. Their exposed trunks and leaves regrow from the underground portion of the plant and so can withstand a great deal of damage. This network of grass stems can also stabilize loose soil and consolidate steep slopes. The grass trees form the basis of a whole new ecosystem.

Back in the Human era, bamboo was the only type of grass which could generate a woody stem. On the Great Plateau, many species of grass have the ability to produce resilient stems. The lower slopes are clad with forests of grass trees. Stems grow out over rocky outcrops and form woody creepers, gnarled and tangled, reaching like fingers towards the next pocket of soil. The photosynthetic part reaches upwards, sending out sprays of leaves from the central trunk. From a distance, a grass tree looks like a cluster of conifers growing from a tangle of rooty creepers.

The oceanward slope of the Great Plateau is green with grass trees. Ridges and banks of vegetation undulate down into the layer of cloud drifting up from the sea. Beyond a narrow coastal plain, sunlight glints on the crest of the waves. In among the rising clouds and shafts of sunlight glide the silhouetted forms of mighty birds.

The birds are great blue windrunners, descendants of Human-era cranes. Long, narrow wings, like those of an albatross, sweep them up towards the distant peaks at high speed, skimming the tops of the grass trees and rocky outcrops. They spend nearly all their lives in flight,

A bird with four wings

The Great Blue Windrunner is an agile, versatile flyer. Throughout evolutionary history, birds have developed different wing shapes according to their flying habits. Long, narrow wings give the best aerodynamic performance in open skies, but short wings are better for maneuvering among trees. Birds that swim underwater would be hampered by large wings, but birds of prey carrying their victims back to the nest need large wings to support the extra weight. By using its feathered legs as extra wings, the great blue windrunner can be a master of many types of flight.

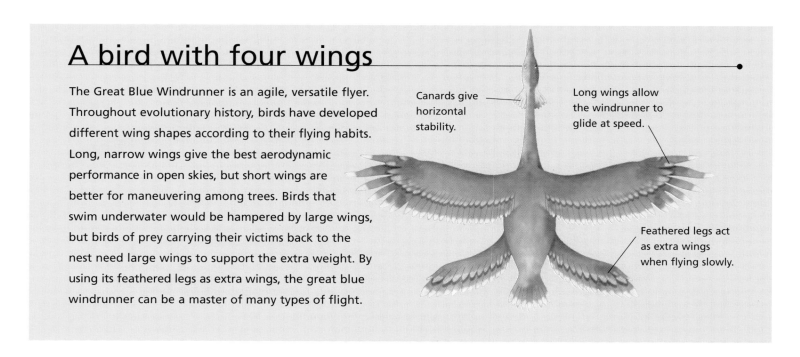

Canards give horizontal stability.

Long wings allow the windrunner to glide at speed.

Feathered legs act as extra wings when flying slowly.

left

The great blue windrunner can soar at high altitudes, making use of its long narrow wings. At low speeds, the bird requires greater maneuverability, and so deploys an additional pair of wings for extra surface area and uplift.

overleaf

A female windrunner tends to her young. The birds nest high in the Great Plateau, far from predators.

soaring and drifting on upthrusts of wind blown up from the sea and deflected from the mountain slopes. The great blue windrunner needs only short periods of rest. As it reaches the highest peaks, it switches to glide and naps for a few minutes, like a Human-era swift.

The windrunner's wings are ideal for covering the great distances between the scattered feeding sites of the Great Plateau. Descending to the sites to feed, on the other hand, calls for low-speed maneuverability. For that, a broad wing is needed. To overcome this obstacle, the windrunner has developed a second set of wings. Feathered and muscled legs, which are normally folded back behind the bird, are brought forward when low speeds are required. When deployed, the wing-legs are held slightly upwards,

improving stability. The bird's flight surface is now increased, providing extra uplift and allowing it to cut its speed and dive with great agility. Additional control is given by canards – feathery eyebrows at each side of the head.

At 33,000 feet (10,000 meters), the thin air provides little protection from ultraviolet light – a serious threat to the great blue windrunner. The bird's metallic-blue coloration helps to reflect ultraviolet light, but its eyes also need shielding. Like all birds, the windrunner has a nicitating membrane, or third eyelid. In addition, the windrunner's eyelid membranes are polarized, forming a pair of natural sunglasses. Adapted in this way, the bird uses ultraviolet light to its advantage. Females identify males by their rich patterns, which are only visible under ultraviolet light.

The great blue windrunner is not the only resident of the Great Plateau to make use of ultraviolet light. Just as some species of spiders in the Human era spun webs with an ultraviolet sheen in order to attract insects, so too do their descendants. Spread across the slopes of the Great Plateau, giant silver webs billow gently in the wind. Their ultraviolet sheen attracts not just insects, but also the unwanted attention of windrunners, which swoop down to feed on the spiders that scurry about busily on the silken fibers.

The humid mountain wind blasts upwards through a gap between the jagged peaks at the edge of the Great Plateau. Fluffy seeds from the grass trees of the lower slopes are swirled upwards in a white blizzard, eddying between the crags. The seeds bristle with tufts that catch the air and act as parachutes. Here and there, they touch the rock, but if there is no soil to engage their little anchors, they are swept onwards once more. From time to time, at one of these landing points, they pick up a little passenger.

A spider, silver in color and measuring only a fraction of an inch across, has grasped one of the seeds and is using its voluminous parachute to carry both of them aloft. This is a silver spider and, as it rides the gusts of wind clinging to the seed, it pays out a strand of silk moored to the rock from which it hitched a ride. Often the seed takes the spider so far that its journey is pointless, but on this occasion the two touch down on the other side of the gorge, where the silver spider tumbles off and secures the other end of the web strand.

Once a silken tightrope bridges the gorge, other spiders appear on the crags and crawl across the tightrope, spinning strands as they go. Soon the gully is spanned by a web, which is continually built up and expanded. The anchor line is strengthened with more strands until it is as thick as a cable. Below it dangle vertical supports, and sheets of fine webbing are spun between these. The structure forms a net across the chasm, spun from up to 15 miles (24 kilometers) of silk and designed to catch insects and seeds swept up by the winds.

left

A silver spider busily spins one of the giant webs that cloak the slopes of the Great Plateau. Grass seeds from the low-lying slopes are blown into the webs.

left

A silver spider scuttles across the web. Her metallic coloring reflects the light, dazzling predators, while the green stripes on her body mimic a grass seed.

A web spun from such a large amount of silk could not be built unless the spiders worked together under a loose social structure. Although not as formalized as that of ants or termites, the social structure in a silver spider colony is nevertheless unique in the history of arachnids. While there is a caste system which dictates the roles of individuals according to their group, as with communal insects, the castes of silver spiders are divided according to age and size. Differently-sized spiders perform different tasks, and as they grow they move into a new caste.

The tiny spiders crossing the gorge on grass seeds are only juveniles. Should they survive the flying ordeal, these youngsters will grow to become the spinners, building intricate traps and barrier webs. Older females are foragers, gathering seeds and insects from the webs and carrying them away to chambers in the clefts and cracks of the rocks. Foragers trim the seeds and carry them in sacs attached to their spinnerets – these females have

forgone breeding and so what were once egg sacs now serve a new purpose. The castes that work on the web have kept their metallic sheen, but have also developed grass-colored stripes. The shiny surface is designed to reflect light and confuse predators. Great blue windrunners patrol these crags, hunting for silver spiders. If they are not dazzled by the spiders' reflective coloring, they may mistake the striped body for a grass seed and lose interest.

Silver spiders are much bigger than their Human-era equivalents. Spiders which reach sexual maturity become the breeders, or queens, of the colony. The queens remain sedentary for much of their lives and can grow to the size of footballs. Unlike eusocial termite colonies, where all the castes serve a single queen, silver spider colonies have multiple queens. A single female could not produce enough eggs to fulfill all the labor needs of the colony. Eggs are laid in spring, when food is plentiful, and are tended by foragers, which also care for the newly-hatched youngsters.

left
A poggle feeds on grass seeds harvested by the silver spiders. This small furry creature is one of the last surviving mammals on Earth.

In the summer, as clouds of seeds sweep up the valleys, the juvenile spiders are released on to the slopes to catch the wind. Some drift away to new locations where they start new colonies. New colonies take time to develop and many fail before their members are mature enough to breed. Any youngsters that are not carried away remain to lay down the foundations of more giant webs, built to catch that season's harvest.

As the web billows in the wind, the foragers swarm over it, disentangling the insects that have been caught there. However, insects make up a very small proportion of the catch. The vast bulk of it consists of the grass seeds that fill the air. All are gathered up and dragged into crevices beneath the mountainside where they are carefully stashed in great seed mounds. But what do these carnivorous spiders want with such large quantities of grass seeds, and why do they exert such effort in gathering them?

In the depths of the colony, the seeds are stacked against the rocky walls, packing the cracks between stones, and piled in loose pyramids. The spiders come and go, adding to the heaps all the time. Suddenly, inside one of the piles, something moves. A little furry face peeps out, all whiskers and rounded ears, big eyes peering around in the darkness. This is a poggle, a small rodent here to feed on the grass seeds.

The time of the mammals is long past. With the extinction of terrestrial dinosaurs, 65 million years before the Human era, mammals flourished. For 150 million years, they were the dominant and most spectacular of the land animals, filling all the evolutionary niches hitherto occupied by reptiles. Now they are fading. The changing temperature and atmosphere of the planet are encouraging the growth of land arthropods: insects, crustaceans and arachnids. The only mammals left are strange, highly specialized creatures.

The poggle is one of these. It lives within the spider colony, feeding on the grain stored with such effort by the silver spiders. The poggle has an easy life. It doesn't need to travel or fight to find food because food is all around it. All it needs to do is eat and breed. Poggles are very fertile and produce large numbers of offspring.

Far from trying to protect their store against the hungry poggles, the silver spiders seem to tolerate the presence of the little mammals. They let them burrow through the piles of seeds, allowing them to feast. But now and again, especially during the spider hatching season, members of the foraging caste descend, search out a fat poggle, seize it and inject it with a paralyzing poison. The twitching body is then dragged into the presence of the queens, where it is left for the poison to start working into the tissues. Enzymes in the poison begin to break down the flesh and before long lumps can be torn from the body to feed the queens, the newly hatched youngsters and the rest of the colony.

The poggles are not just tolerated, they are actively encouraged. The carnivorous silver spiders have no interest in harvesting the seeds for their own food, they are farmers of a different kind. The grain harvest is there to feed up the livestock before it is butchered.

The poggle is essential to the survival of the silver spider colony. An abundant supply of meat gives the colony a far greater chance of making it through the harsh winter. But it is not just proteins from the poggles' flesh that are harvested. Live pregnant females are taken to the queen spiders, which drink their blood. Hormones generated by pregnant poggles actually stimulate the egg production of the queens. The remaining flesh then goes to feed the rest of the colony.

The last representatives of the mammal dynasty may have an easy life, free from the cares of finding shelter and food, but in death they serve as nourishment for the new masters of the Great Plateau.

right

A queen silver spider squares up to her next meal. Hormones from the poggle's blood will stimulate the queen's reproductive system.

MASS EXTINCTION

A BLACK CLOUD REACHES ACROSS THE BLUE SKY of the world, 100 million years after the Human era. It is generated by volcanic convulsions in the subduction zone between the continents of Asia and Antarctica. Tall, jagged volcanoes erupt in huge explosions, sending pyroclastic flows – mixtures of gas and superheated rock – racing down the mountain slopes, incinerating everything in their path.

More volcanic activity blasts out along the western edge of the Asian continent and the northern edge of Africa. There, basaltic volcanoes throw up a continuous stream of black lava that creeps over the land. More and more lava spews from the mountains, spreading layer upon layer of sterile black rock across Earth until, over thousands of years, huge areas are covered.

Gas, dust, ash and water vapor are belched out of the volcanoes and gather in the atmosphere. There they are picked up by the prevailing winds and spread westwards along the equator. Eventually they are swirled north and south, spreading as far as the poles. Acids in the gases combine with the water vapor and fall as acid rain. The upper atmosphere is filled with fine particles of dust and ash, blocking the incoming solar radiation. Black clouds form a pall across the sky.

In the gloom, global temperatures plummet and animals begin to die. Plants, deprived of sunlight, shrivel up. Even the hardiest creatures starve. A few lost links in a food web have repercussions throughout the system. Whole communities wither and die away.

The temperature of the Shallow Seas falls and the prevailing currents fail. Ocean phantoms deflate and die. Reef gliders shrivel for lack of food. In the Bengal Swamp, plants wilt and are incinerated by blasts of white-hot ash, while cold-blooded toratons stand motionless, their vitality seeping away in the cold. On the Antarctic continent, flutterbirds choke in the poisonous atmosphere. High on the Great Plateau, the webs of the silver spiders hang limp and torn, deprived of their seasonal influx of seeds. Great blue windrunners are grounded and flap helplessly in the face of storms generated by the climatic disruption. It looks as if all life is doomed. Yet, several times in the geological history of Earth, mass extinctions such as this have swept the globe. And life has survived.

"Volcanic activity, both under the sea and on land, characterizes this hothouse Earth, 100 million years after humankind. As a result of all the cataclysmic activity, there is a great increase in the amount of carbon dioxide, toxic gases and dust being thrown into the atmosphere. The resulting acid rain leads to the widespread extinction of plants and animals."

Dr Roy Livermore
Paleogeographer
British Antarctic Survey, UK

right

It is the end of an era. A long period of warm, stable conditions is brought to a close with a burst of volcanic activity. The atmosphere is filled with dust and ash, cutting out the sun and leading to a mass extinction.

200
MILLION
YEARS

A NEW PANGAEA

"We can be reasonably confident that, 200 million years in the future, the continents will once again come together to form one supercontinent, surrounded by one ocean. The formation of supercontinents tends to have an enormous influence upon the weather patterns of the globe. It creates a globe of extremes."

Professor Bruce Tiffney
Paleobotanist
University of California

THE PLANET HAS CHANGED. It is 100 million years since the mass extinction that destroyed most of the life on Earth. The continents have continued to move, driven by the relentless processes of plate tectonics. Now, 200 million years after the Human era, Earth presents quite a different face.

The world has come full circle. For many millions of years, spanning much of the Permian and Triassic periods, Earth was one landmass, termed Pangaea. Now, hundreds of millions of years later, the continents have once again come together, forming a single, huge supercontinent: Pangaea II. The supercontinent covers most of the northern hemisphere, from what was once the North Pole down to the equator. It is a place of hostile deserts, sprawling mountain chains and sodden coastlines, whipped by constant hurricanes.

Surrounding Pangaea II, there is nothing but water, a Global Ocean that governs the extreme climates of the era. The pull of the Moon has gradually slowed the rotation of Earth and a day is now 25 hours long. With more sunlight to warm the surface water, violent storms develop, battering the coastal regions.

Even these mighty storms cannot carry moisture far into the supercontinent. Inland, thousands of miles from the ocean, lies the desolate terrain of the Central Desert. Here, average temperatures range from over 120°F (50°C) in the summer to -20°F (-30°C) in winter. The only water comes from springs rising up from a massive subterranean ecosystem of caves laid down when this region was covered by shallow seas, 100 million years before.

At the southern edge of the continent, a towering, volcanic mountain range forms an impenetrable barrier against the saturated onshore winds. The moisture rich clouds rising over the peaks lose water as they cross, starving the land behind of nourishment. This is the Rainshadow Desert, where life ekes out an existence from whatever residual moisture is provided by the storm-drenched coast.

On the north coast, there are no such defenses. The area is pounded by westerly storms, bringing large amounts of water. The constant rain and warm climate have created a hothouse environment, and a vibrant, vigorous forest has sprung up, teeming with life. This region is called the Northern Forest.

It is a tough world, but still it is full of life. The planet needs around 10 million years to recover from a mass extinction and 100 million years have now passed since the last one. The survivors have branched out, diversified and established themselves in new and surprising evolutionary niches.

Pangaea II, the new supercontinent

Slowly, the great continental masses of the Human era have become one, forming a giant supercontinent, or Pangaea II. What was once Africa has turned and fused to the southernmost tip of Asia. Australia long ago moved north and collided with the southeastern edge of Asia, and Antarctica subsequently followed it into the gap. North and South America have swung to the east, squeezing the oceans out of existence, and sealing in the rest of the landmass.

Earth today
The separate continents and oceans provide diverse climates and habitats for evolution, which means that species evolve individually.

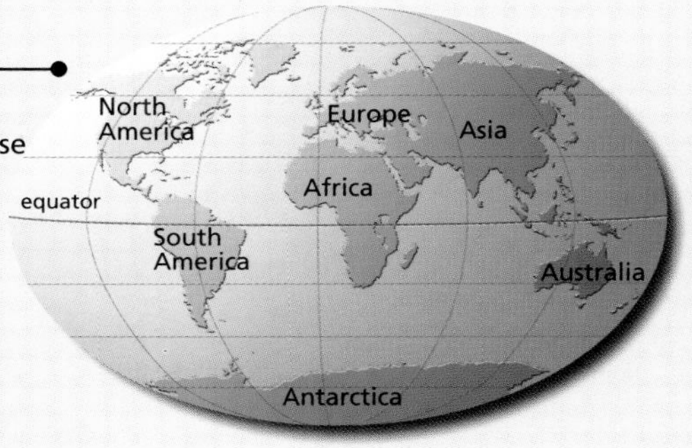

North America   Europe   Asia

equator   Africa

South America

Australia

Antarctica

Pangaea II
A single landmass surrounded by a Global Ocean makes this a world of extreme climates. An immense current, involving the whole ocean, circulates in an anticlockwise direction around the southern hemisphere. This global current makes it easy for life to migrate around the ocean.

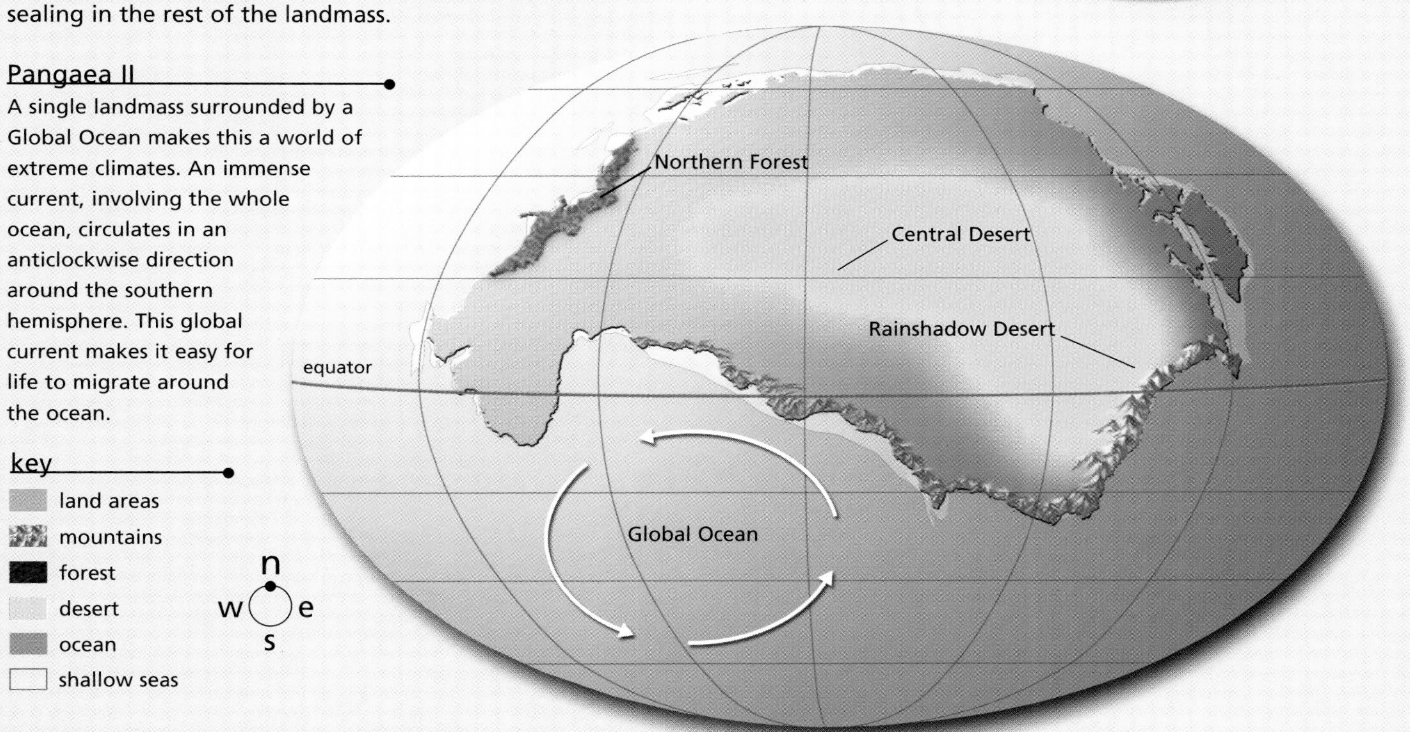

key
land areas
mountains
forest
desert
ocean
shallow seas

n
w   e
s

Northern Forest

Central Desert

Rainshadow Desert

equator

Global Ocean

111

the CENTRAL DESERT

IMAGINE THE DESERTS OF THE HUMAN ERA. The boundless continental deserts, such as the Gobi and the Kyzyl Kum, which are so distant from the sea that little moisture reaches them. The rainshadow deserts, such as Death Valley, which are tucked in behind coastal mountain ranges that form a rocky barrier to the moist oceanic air. The so-called trade wind deserts, such as the Sahara, where clouds are chased away by hot winds, leaving the sun to scorch the sand. And the coastal deserts, such as the Namib, where cold ocean currents cool the air and desiccate the land. 200 million years after humankind, these deserts have run into one another. Separated only by jagged mountain peaks, they cover the largest continent that has ever existed.

The Central Desert is a wilderness of drifting sand-seas, sun-cracked stones and shattered gravel. Unbearable extremes of heat and dryness combine to produce the most hostile living conditions. There are no clouds and the summer sun sears the bare rocks and sand so that temperatures reach a withering 120°F (50°C) in the daytime. At night, the accumulated heat is radiated away to the frosty sky and temperatures dive to a bitter -20°F (-30°C). In winter, parts of the northern interior are colder than Earth's surface has ever been.

Such are the conditions for life – or survival – at the heart of Pangaea II. The most remote areas of the Central Desert have not seen rain for hundreds of years. So where is the water that is essential for life? This region was once covered by warm, shallow seas, formed when sea levels were high and the climate was temperate. Gradually, as the continents piled into one another, the land was uplifted and the shallow seas drained off into deep ocean basins. Rain filtered into the limestone and created a sprawling labyrinth of limestone caves deep below the Central Desert.

At the edge of Pangaea II, constant rain drenches the seaward mountain slopes and soaks into the strata, eventually seeping into the porous limestone of the mainland. Over time, this water fills the subterranean reservoirs that lie below the Central Desert, giving life to the barren wastes above. The animals and plants that exist in this arid land are true specialists. Experts in enduring extremes of temperature, they survive through the single-minded pursuit of water.

"On the supercontinent, 200 million years in the future, there will be an incredibly dry desert. The center of this desert will be thousands of miles away from the nearest water mass. This means that it will not only be very dry, but there will also be extreme swings in temperature. The land will warm up and cool down very rapidly, causing massive temperature variations between night and day, and between the seasons."

Professor Paul Valdes
Paleoclimatologist
Reading University, UK

left

The surface of the Central Desert is an inhospitable environment, but life has adapted to survive the extreme temperatures and lack of water.

The most successful living creatures in the Central Desert are insects. For 600 million years, their remarkable adaptability has enabled them to survive the most extreme conditions and weather out all the great mass extinctions the planet has suffered. More than any other living organism, insects are able to diversify into and exploit any number of ecological niches.

Insects have many different strategies for survival. Because of their size, they can colonize the smallest crevices and shelters. They are covered by an impermeable cuticle that protects their internal organs in a watertight armor, enabling them to exist in even the driest of places. Their capacity for flight means that they can disperse to new localities when conditions become tough. Also, their lifecycle is often divided into separate phases: a larval phase and an adult phase. In this way, larvae and adults have different food habits, so competition between them is reduced and they are more likely to survive.

In the hostile environment of the Central Desert, insects have found a way to create their own living conditions. Here and there on the exposed surface, jagged, dusty towers rise up out of the sand, jutting many feet into the air. At the pinnacle of each structure there are translucent panels that could almost be windows. These towers are the cities of the terabytes – the most common insect in the desert.

Terabytes have evolved from termites. Like their ancestors, terabytes are eusocial, which means they form a colony in which a single queen breeds while every other individual performs distinct tasks essential to the colony's survival. In colonies like these, individuals are divided into several castes, each with a function of its own.

The fiercest terabytes are those in the warrior caste. Their disproportionately large heads generate chemicals which they squirt from nozzles to fight enemies and secure prey.

left

Each terabyte caste fulfills a distinct and essential role within the colony. Here, a transporter carries a warrior in search of prey. The warrior squirts chemicals at its victim through a nozzle in its head.

right

The strange towers dotted about the desert are built by terabytes in a sophisticated feat of architecture and engineering. Within each tower they breed, mine water and farm algae for food.

From time to time, the terabyte farmers must renew their stocks of algae, no easy task when vegetation is scarce. Terabytes solve this problem by removing algae from the tissue of another animal.

At daybreak, the garden worm comes slithering out of fissures in the rock and spreads itself out to catch the rays of the rising sun. It humps up its middle section and begins to unfurl green, fern-like tissues that branch and fan out from the segments of its body. These 'leaves' are fleshy extensions of the garden worm's body, but they can act like real leaves because they are packed full of algae. Just like real leaves, they convert sunlight into food.

The garden worm is a polychaete, a type of segmented worm that has been highly successful throughout the history of life on Earth. In the Human era, polychaetes, also known as bristleworms, were among the most common marine organisms, numbering around 8,000 different species. They adapted to life in a variety of habitats, swimming freely in the deep ocean or the surface waters or burrowing in the mud and sand of the beach.

About 18 inches (45 centimeters) long and some 1.5 inches (3.5 centimeters) high, the garden worm's natural habitat is the maze of flooded crevices and caverns riddling the limestone beneath the Central Desert. But food is scarce here and so the garden worm has solved the problem by developing a symbiotic relationship with the green algae. The algae provide nourishment, the worm provides locomotion, carrying the algae to the surface sunlight in order to photosynthesize.

Sunbathing is a risky business, however. A party of terabyte warriors, carried by transporters, has spotted the basking garden worm. They swarm over the rocks, the warriors firing sticky threads from chemical nozzles on the heads of the warriors, gluing the unwary worm to the rock. Transporters begin to cut away at the immobilized worm's algal tissues. The garden worm fights back. It issues a secretion that dissolves the glue. The bonds weaken and tear and soon the worm is free. It furls its green lobes and wriggles away to safety. The battle is over, but the terabytes have been able to tear off enough algae-rich tissue to restock their greenhouses.

Convoluta worms

Many animals have learned to live symbiotically with algae. Convoluta worms, which lived in large colonies on the beaches of Europe in the Human era, were dark green in color because of the algae living within them. At low tide, during daylight hours, they came to the surface, forming green patches on the sand. They had no stomach and needed to absorb food directly, so they basked in the sunlight in order to grow their own food, allowing the green cells inside their bodies to photosynthesize. In return, the algae received nutrients derived from the worms' bodily waste.

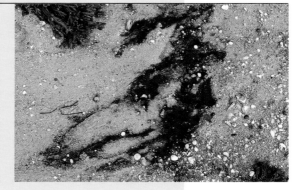

Human-era Convoluta worms sunbathe on a beach. Their green coloring comes from the green algae which they carry inside their bodies.

right

The garden worm basks in the sun, allowing millions of algae that grow on its back to photosynthesize.

The sunburnt surface of the Central Desert belies the labyrinth of limestone caves and water-filled fissures beneath. To explore this hidden world, it is necessary to look in more detail at its formation. The limestone deposits are a relic of the reefs and muds of the shallow seas that once existed here. As the continents collided, the land rose up, displacing the shallow seas and compressing the muds and reefs into solid stone. Limestone is made up of the shelly debris of marine life and is easily eroded and dissolved by acids in the groundwater. The action of these acids initially creates small pores but, over time, large caves can result. Despite the acidity of the water and the total lack of sunlight, one group of animals has flourished: polychaetes, a family of segmented worms also known as bristleworms.

Several species of worm live below the Central Desert and they are all descended from a single ancestor, a marine polychaete that thrived in this region when it was covered by shallow seas. When sea levels dropped, the worms were trapped, but they survived and adapted to their new environment, slowly evolving to fill the different niches the cave system offered. This process is called adaptive radiation.

Normally all life derives indirectly from sunlight: plants convert carbon dioxide and water to food using the energy of the sun; plant-eating animals eat this food and are in turn eaten by carnivorous predators. In the darkness of the caverns, this does not apply. Instead, the initial energy is derived from chemicals. Bacteria break down sulfur compounds in the rocks and grow on the energy released, forming an encrustation on the cave walls.

The green bacterial meadows are grazed by the gloomworm, one of the surviving polychaetes. The gloomworm lives above the water level where it can avoid larger predators.

The subterranean reservoirs of the Central Desert are the hunting ground of one such predator. The slickribbon is another member of the polychaete family. It is about three feet (a meter) long and swims by means of paddles, one pair to each segment, beating in a wave-like action like a kind of aquatic millipede. The slickribbon's mouth parts are mounted on an extendible trunk, which it can shoot out in a fraction of a second. Gloomworms that come too close to the surface of the water are its staple food source.

Garden worms are also hunted by the slickribbon when they retire to the caverns at night. They are more nutritious than gloomworms but more difficult to catch, dodging the hunters with their agile swimming action. The struggle for survival beneath the Central Desert is as constant as that above it.

left

The slickribbon is a fearsome predator. Its jaws are mounted on an extendible trunk that snaps out at passing prey.

right

A slickribbon catches a gloomworm in its powerful jaws. Gloomworms lingering near the surface of the water are a tasty food source for the slickribbon.

the CENTRAL DESERT

the GLOBAL OCEAN

THE MASS EXTINCTION OF 100 MILLION YEARS after humankind didn't just ravage life on land, life in the oceans was profoundly affected too. Active volcanoes filled the sky with ash and dust, cutting out the sunlight for months on end. Acid rain, formed by sulfur compounds belched out by volcanoes, fell continually into the sea.

The lack of sunlight and the increase in acidity killed off the plankton in the surface waters and led to a catastrophic collapse in the oceanic food chain. Bony fish – the dominant marine animals for hundreds of millions of years – suddenly died away, along with all kinds of other creatures. Where once the oceans had teemed with life, they became almost barren. But nature does not leave ecological niches vacant for long. The animals that survived the mass extinction did so because they took shelter in the deepest, most remote refuges of the ocean. Once conditions had stabilized, fish and their relatives were replaced by completely new forms of life.

It is now 200 million years since humans lived on Earth. The planet is dominated by a single, giant landmass called Pangaea II. One continent means one ocean, the Global Ocean, a body of water so vast that its center lies 10,000 miles (16,000 kilometers) from the nearest coast. This uninterrupted expanse of water helps to determine the extreme weather conditions of the planet. The intense heating of the atmosphere at the equator draws in trade winds from the north and south. These converge and blow westwards along the equator, driving permanent ocean currents before them. The result is a constant equatorial gyre – an immense circulatory current that involves the whole ocean (see map, page 111). The global current makes it easy for sea life to migrate, and so the Global Ocean is populated by very cosmopolitan groups of animals.

As the predominant ocean currents run east to west, there is little water movement between north and south. The cold waters of the South Pole do not mix with the warm waters of the equator. The result is a steep temperature gradient between high and low latitudes. However, worldwide temperatures are still too high for there to be a polar icecap. This single ocean is a complex environment supporting intricate food chains and highly-evolved species, quite unlike anything known from the Human era.

"The sea is a soup of tiny single-celled plants that live their lives very quickly. If the sunlight is shut off for a day, then a generation of these plants is lost. If it's shut off for 30 days, that is 30 generations. 30 generations without sunlight could collapse the biosphere in the oceans, reducing all the planktonic productivity to nothing."

Professor Stephen Palumbi
Biologist, Harvard University

left
Pangaea II is surrounded by an ocean so vast that its center is 10,000 miles (16,000 kilometers) from the nearest coast.

the GLOBAL OCEAN

Consider the marine arthropods of the Human era – crabs, lobsters and shrimps. They were all specialized, preferring a particular food source and habitat. However, their larval forms tended to be generalized, subsisting on diverse food sources and under almost any conditions. Arthropods were incredibly prolific throughout the oceans of the world and in some cases, despite being so small as to be invisible to the naked human eye, were effective predators. Versatile and hardy creatures such as these were ideally positioned to adapt and diversify in a time of crisis.

Immediately after the mass extinction, 100 million years after humans, arthropods developed a new and crucial biological ability. They became able to reproduce while still in a juvenile, or larval, form – a phenomenon known as neoteny. Sidestepping the need to develop into cumbersome adults provided them with an evolutionary boost. It allowed them to evolve in diverse ways, filling many of the niches

left by the now extinct bony fish. Soon the oceans were repopulated with a whole family of newly evolved animals. These arthropod descendants are called silverswimmers.

There are almost as many different species of silverswimmers in the Global Ocean as there were fish in the Human era. They all have a similar body plan – a lightweight armored head with bristly legs and antennae protruding from beneath, and a segmented tail that drives the animal through the water with an up-and-down motion. Like fish, they have branched out to take advantage of every habitat and opportunity. There are flat, bottom-dwelling silverswimmers; fierce, hunting silverswimmers; large plankton-feeding silverswimmers; and even silverswimmers that live as parasites on other silverswimmers. They range from the microscopic to those the size of a small whale. They are the success story of the Global Ocean that surrounds Pangaea II.

left

Silverswimmers are the most prolific species in the Global Ocean. They are descended from crustaceans, such as crabs and lobsters.

right

This species of silverswimmer roams the ocean in shoals. A complex array of antennae and bristles at the front of the animal enables it to filter out fine particles of plankton from the water and sweep them into its mouth.

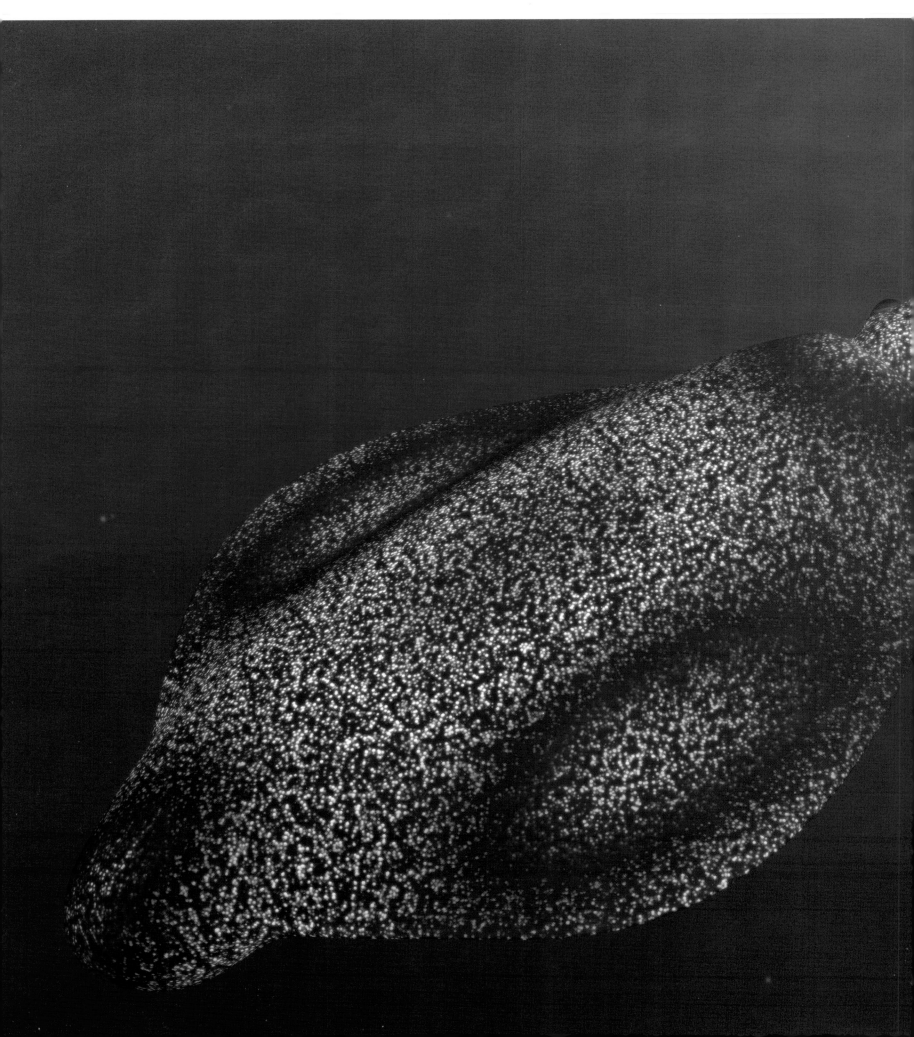

the GLOBAL OCEAN

Huge as they are, rainbow squids still have a few enemies, chief among them an ancient evolutionary success story. Sharks have been the hunters of the oceans since Devonian times, 400 million years before the Human era. They are survivors, sleek and efficient, and have outlived all the mass extinctions that have occurred, adapting to whatever new environment presents itself. The simple, primitive design of the shark and its brutally straightforward lifestyle have ensured its survival.

With the formation of Pangaea II and the resulting vast ocean mass, there are new problems for these marine hunters. Shark food is now widely dispersed, with huge stretches of empty water between. A single shark hunting randomly is unlikely to come across enough prey to sustain it and may be unlucky enough to starve. But many sharks hunting over a large area are in a better position to meet a food source. When they do, if they are able to communicate the presence of food to the others, the whole shark population benefits. The sharkopath has evolved just such a method of communication.

The Global Ocean is patrolled by loose groupings of sharkopaths covering a large expanse of water. If one happens upon a rainbow squid, it sets off a flashing sequence in bioluminescent patches along its side. This visual signal penetrates the water and can be picked up by the sharkopath's closest neighbor. The neighbor repeats the signaling process and soon the whole group of sharkopaths is aware of the presence of food and starts to home in on its quarry.

The rainbow squid's camouflage is so effective that it melts into the background as soon as it notices the danger. By spreading a light color across its belly, criss-crossed with rippling patterns like surface waves, it disguises itself from attackers coming from below. By darkening its back to merge with the depths, it hides from attackers above. Despite this remarkable camouflage, a marauding group of sharkopaths – armed with an incredible array of senses, and with the added advantage of strength in numbers – stands every chance of overcoming the squid's defenses.

left

The spined pygmy shark inhabited the deep oceans of the Human era. Bioluminescent patches on its skin were used to confuse predators.

right

The sharkopath has evolved bioluminescence as an effective means of communication. The bright patches on its skin guide other sharkopaths towards prey.

the RAINSHADOW DESERT

LIKE A GIANT SPINNING TOP losing its momentum, Earth's rotation has gradually slowed. 200 million years after the Human era, this imperceptible deceleration has added an hour to Earth's cycle and a day is now 25 hours long. Furthermore, the sun's luminosity has increased, leading to a rise in air temperature. As the hot sun beats down on the surface waters, it warms up the ocean, leading to frequent and very strong hurricanes.

The east coast of Pangaea II is constantly battered by these powerful storms, or hypercanes. Winds gusting at over 250 miles per hour (400 kilometers per hour) blast the shoreline. As the warm surface waters fuel the growing storms, the swell gathers into huge waves on the narrow continental shelf. The resulting breakers can be over 70 feet (20 meters) high. Meanwhile, rain falls in unrelenting torrents from dark clouds.

The violent winds, and the humidity they bring, do not travel far inland. Along the eastern edge of the supercontinent runs a high coastal mountain range, not unlike the South American Andes of the Human era, but much longer and a good 10 per cent higher. Winds sweep up the seaward faces of the mountains, dropping all their moisture as they rise. As they reach the summits, they spill over, blow through the passes and valleys, and descend as dry gusts into the arid hinterland. Here, starved of moisture by its own rocky boundaries, lies the Rainshadow Desert.

Despite the devastation they wreak, the hypercanes not only provide the Rainshadow Desert with residual humidity, they also supply nourishment in the form of marine life. Animals swimming near the surface of the sea are plucked up by the suction of the hypercanes. The updrafts can sweep even quite large animals a few miles up into the atmosphere. As the energy of the wind is absorbed by friction against the mountain slopes, the blasts abate and the sea creatures fall as organic debris beyond the mountains and into the desert. There is a constant influx of organic matter here, whether it be plankton caught up in the spume blown from the crests of the waves, or the bodies of hapless flish, carried far from their ocean home and dumped unceremoniously in the sand. In a landscape such as this, no food source goes unexploited.

"The hurricanes which smash into the east coast of the supercontinent will be huge. Imagine the sort of storms that hit the Florida coast in the Human era and add 50 per cent to the strength of the wind, the rain and the sheer battering power. These hypercanes will bring tremendous devastation to the coastal region."

Professor Paul Valdes
Paleoclimatologist
Reading University, UK

left

The Rainshadow Desert lies in the shadow of the storm clouds that gather over the coastal mountain peaks. Heavy storms batter the shoreline, but moisture is prevented from reaching the inland desert by rocky barriers.

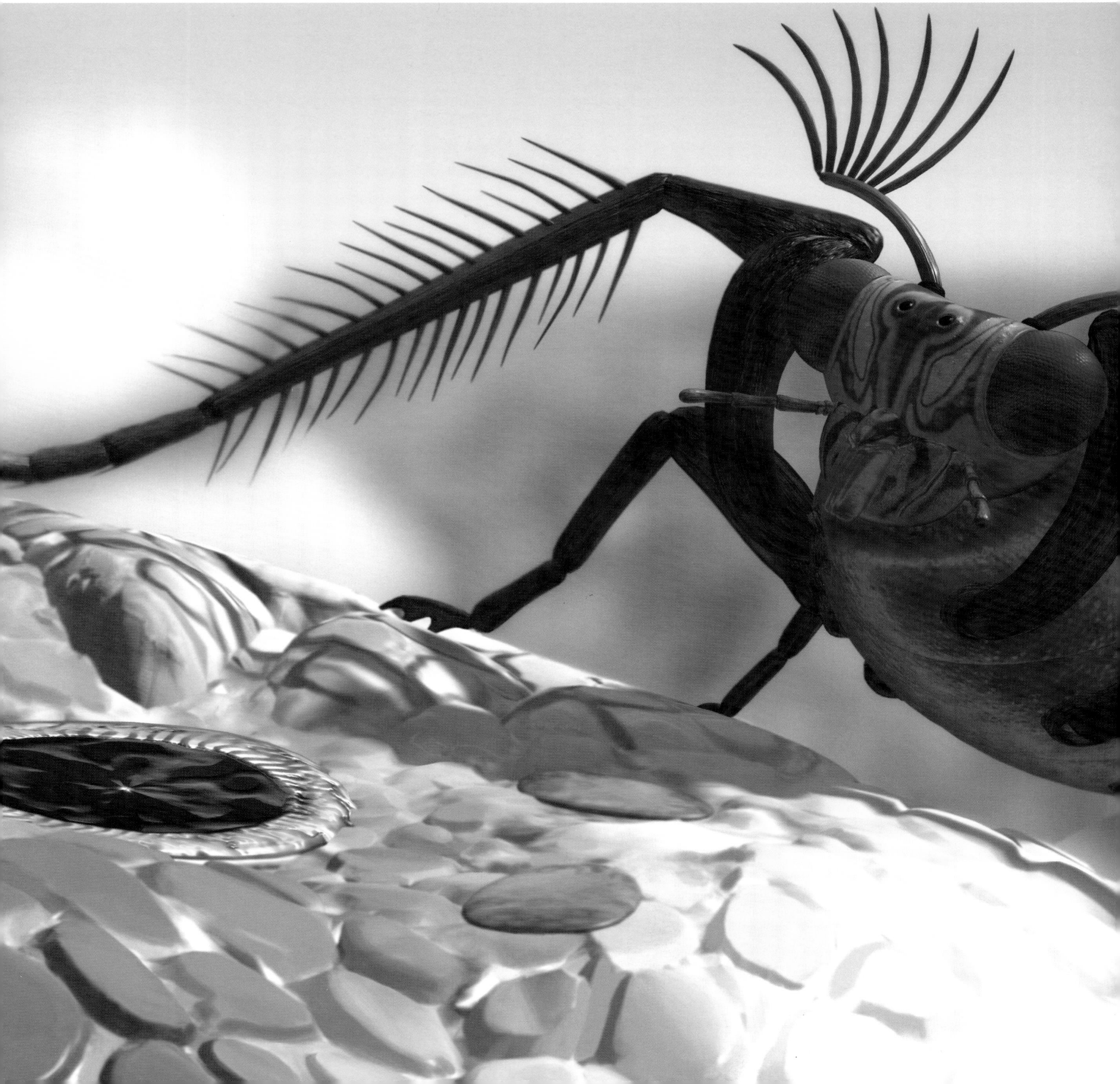

there is a movement in the sand. Something is pushing its way to the surface. First to emerge is what looks like the shell of a large snail. Then the rest of the creature appears, slowly pushing itself from the sand with a muscular foot. Finally, it raises itself, balancing on the tip of the foot, its body forming an elegant, vertical S-shape.

The tip of the foot has three toe-like projections, which spread its owner's weight across the sand. Instead of being mounted on stalks, this animal's eyes are in circular turrets, like those of a chameleon. It lifts its shell clear of the ground and sways for a moment. Then the snail jumps away!

Throughout their evolutionary history, land snails have been slow-moving, rather passive animals. They have always preferred moist places where the plants upon which they feed are abundant. Crawling slowly on a fleshy foot lubricated by a trail of slime, the Human-era land snail rasped food from soft plants using its toothed tongue, or radula. In the choked and unyielding habitat of the

Rainshadow Desert, the desert hopper has evolved to become a successful grazer among the region's sparse, tough plants.

The desert hopper is one foot (30 centimeters) tall. When withdrawn into its 8 inch (20 centimeter) shell, it has few enemies. It does not secrete a trail of slime – a waste of water in a desert habitat. The soft skin of its ancestors has evolved into horny, interlocking scales, forming a tough, lizard-like skin that locks in essential moisture.

Most remarkable of all is the desert hopper's muscular foot. The foot is a jumping organ. It carries the snail across the desert at a speed similar to that of a human jogging. In the Human era, a species of marine cone snail could use its muscular foot to hop away from predators. 200 million years later, this has developed into a method of moving swiftly on land. The jumping action also works a bellows mechanism within the shell of the desert hopper, allowing air to be pumped in and out of the lungs and providing sufficient oxygen for its active lifestyle.

Hopping mollusks

In the Human era, land snails crawled along by secreting slime from the surface of the foot and moving in tiny muscular waves that propelled them across it. However, one species of marine snail modified the foot into a jumping organ capable of launching it out of the sand and away from predators.

In the dry habitat of the Rainshadow Desert, where moisture is a scarce resource, the desert hopper has developed the ability to hop rather than slide on a trail of mucus. Its hard, protective skin prevents water loss and, by hopping, it minimizes contact with the searing heat of the desert sand.

A marine cone snail found in tropical waters during the Human era could hop away from predators on its muscular foot.

The desert hopper's diet consists of tough, fibrous vegetation. The muscular, tooth-covered mouthparts of the Human-era land snail have evolved into a rod-like structure with a toothed drill at the end. This can pierce the waterproof cuticle of a plant to reach the pulp inside. The desert hopper's shell and thick skin provide an efficient defense against most plant thorns and spines. Like many desert animals, the hopper does not drink. It obtains all its water from food, and is perfectly adapted to keep all that moisture within its body.

Despite being a voracious plant-eater, the desert hopper itself is food for a plant. The death-bottle plant is carnivorous, relying on animals to provide the nutrients lacking in the desert soil. Part of the plant consists of an underground chamber, covered by a thin membrane. Overground, a spray of stems and leaves photosynthesize sunlight and act as bait for herbivorous desert animals.

When a hungry desert hopper lands on the thin membrane of the death-bottle plant, hoping to feed on its leaves, the membrane breaks and the snail falls through. Trapped inside the underground chamber, the desert hopper begins to struggle. As it attempts to escape, poisonous spines drive into the exposed parts of its body, piercing its hard skin. The poison subdues the hopper and the plant's digestive juices start to break down the animal's tissues. After a few days, the hopper has been fully digested and the membrane has regrown, ready for the next victim.

There is only a sparse population of death-bottle plants in the desert so they have had to develop a failsafe strategy for propagation. Most plants reproduce by cross-pollination, using the wind or airborne animals to disperse male cells, in the form of pollen, to the female parts of other plants of the same species. The death-bottle plant cannot afford to take such risks. Its pollen might not reach a female plant, so fertilization and seed production might never occur. It has therefore developed a strategy of self-pollination, whereby pollen is transferred to the female part within the same plant, allowing it to produce seeds. It is to disperse these seeds that the death-bottle plant calls on the services of the bumblebeetle.

The plant has developed a leaf with the shape and silvery sheen of a dead ocean flish. Enzymes in the leaf can even recreate a convincing odor of rotting flesh. A passing bumblebeetle lands on a flish-like leaf to deposit its grimworms but, burrowing into the leaf, it too finds itself in a trap. Unlike the spiny chamber the desert hopper was caught in, however, this chamber is full of sticky seeds.

The bumblebeetle blunders around in the underground chamber until, suddenly, a spring mechanism in the plant catapults it skywards. Seed-coated but safe, it flies away, continuing its quest for a real flish. As it lives out its short life, flying towards a flishwreck or fighting for possession of a dead flish, the heavy seeds fall from the insect. From these seeds, if conditions are right, new death-bottle plants will grow.

The relationship between the bumblebeetle, the flish and the death-bottle plant is yet another example of how, when faced with an environment as harsh as the Rainshadow Desert, organisms can adapt to make brilliant use of each other.

right
Once the desert hopper is trapped in the digestion chamber of the death-bottle plant there is no escape.

Low over the scrub of the Rainshadow Desert comes a bulbous shape on flashing wings, roughly the size of a Human-era sparrow. No sooner does it appear than it vanishes again. Its swift passage leaves a flicker of disturbance in the sparse desert grass, and a thrumming sound hanging in the air. It is a bumblebeetle, an insect inhabitant of the dry regions of Pangaea II.

The teardrop shape of its body causes little air resistance as it skims along, dipping over sand dunes and weaving in and out of the hardy vegetation, constantly navigating its way clear of obstacles. Its membranous hind wings drive it onward, while the hard forewings, called elytra, are spread out to form an aerodynamic surface. These natural airfoils provide lift, allowing the insect to cut off its power and glide from time to time. By conserving energy in this way, it can cover great distances without tiring – even though this is the only flight it will take in its life.

The stiff elytra are more than airfoils. They also act as supports for sophisticated sensory apparatus. The beetle's forelegs, coated with scent receptors, are held along the leading edge of the forewings. Additional sensors dangle from the tail. Indeed, every available surface of the bumblebeetle is covered in sensing hairs. This array is necessary, as the food sources the insect is hunting are spread out across the desert.

About 40 per cent of the bumblebeetle's streamlined, yet bulbous body consists of fat – a store of energy built up during its larval stage. This will fuel up to a day of continuous flying, in which time the bumblebeetle can cover some 500 miles (800 kilometers). At some point in the journey it is likely to sense food and close in on it.

left
At the end of its flight, the bumblebeetle settles on a dead flish, ready to spread its larvae into the nutritious mass.

137

The food this insect seeks is a dead flish. Large numbers of ocean flish are carried over the mountains from the Global Ocean with every hypercane. By the time their carcasses tumble into the dust of the Rainshadow Desert, they have been battered and desiccated by the ferocious winds. Here and there across the arid landscape lie 'flishwrecks' – whole shoals dropped in a small area.

A flishwreck is a valuable source of nourishment to all kinds of desert-dwelling animals. The bumblebeetle, however, has evolved to rely exclusively on this food supply. Its sensory array can detect a mass of rotting flish miles away, and each individual bumblebeetle will fight its own kind to take possession of a carcass. Only one bumblebeetle can secure each flish.

The scent of decaying flish spreads downwind, often close to the ground, and any insect lucky enough to detect it begins to home in. Before long, the air is full of the buzz of bumblebeetles, each seeking an unclaimed carcass. If two insects approach an unclaimed flish, a fierce aerial skirmish breaks out. These fights are rarely fatal but the loser must relinquish the prize and find a carcass of its own. Failure to find a flish means death not only for the bumblebeetle, but also for the load it is carrying.

When a bumblebeetle claims a flish, its quest is over and so is its life. The hard-earned food is only for its offspring. Indeed, it could not enjoy a flish supper if it wanted to: the adult has no mouthparts and only the most rudimentary digestive system. Its body consists of wing muscles, a fat supply – and its young. As the insect alights upon the dead flish, its abdomen collapses and it dies. In doing so, it releases the larvae into their new home.

The bumblebeetle has lived for a day and fulfilled its sole purpose – to find a feeding ground for its offspring. If it does not succeed within a few hours, its fat supply runs low and it begins to digest the larvae it carries in its abdomen. Individuals are sacrificed in order to maintain the genetic line. If a single larva survives to be placed in a flish carcass, then the bumblebeetle has done its duty.

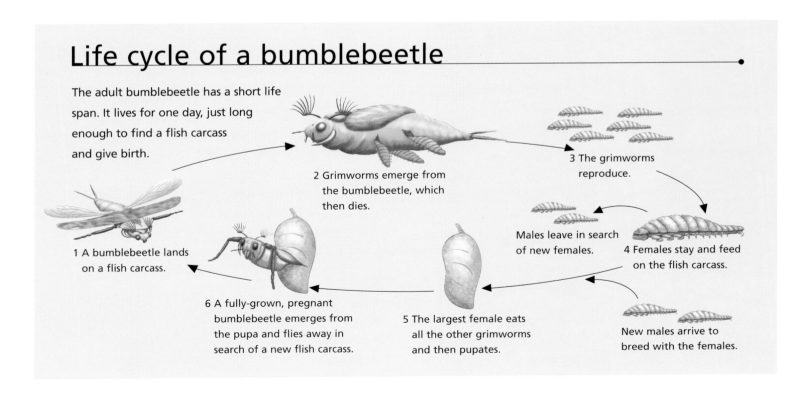

Life cycle of a bumblebeetle

The adult bumblebeetle has a short life span. It lives for one day, just long enough to find a flish carcass and give birth.

1 A bumblebeetle lands on a flish carcass.

2 Grimworms emerge from the bumblebeetle, which then dies.

3 The grimworms reproduce.

Males leave in search of new females.

4 Females stay and feed on the flish carcass.

New males arrive to breed with the females.

5 The largest female eats all the other grimworms and then pupates.

6 A fully-grown, pregnant bumblebeetle emerges from the pupa and flies away in search of a new flish carcass.

left

Grimworms emerge from the body of the bumblebeetle.

The larvae, called grimworms, spill out from the adult's body and burrow their way into the decomposing flish. If the adult bumblebeetle lacks mouth parts, its offspring certainly make up for it. Inside the sucking, disc-shaped mouth of the grimworm is a terrifying mill of sharp teeth. These tear through the flesh and bones of the flish carcass and suck up its putrefying internal organs.

Narrow tunnels are bored into the flesh. These are then enlarged into chambers as the grimworms chew away at their new home. The only part of the flish left untouched is the skin. Under the desert sun it shrivels to a hard leathery case, providing an ideal shelter for the grimworms. Without it they could not survive the searing heat and scouring desert wind. Safe in their flishy capsule, they live, feed and grow.

To ensure the full exploitation of each flish, grimworms reproduce by parthenogenesis, which means that unfertilized eggs develop directly into identical larvae.

This process continues until what remains of the flish's body has become a writhing mass of grimworms. Some of these individuals are male; others are female. As they feast, they grow. Eventually, one of the females reaches its optimum weight of around half an ounce (10 or 15 grams), enormous for an insect larva. The female grimworm remains in the flish and continues to devour it along with her smaller siblings. The males burst out, chewing through their leathery tent, and set off to find another, nearby flish.

In a typical flishwreck, each corpse is a short distance from the next and all are at the same state of decay. They are also all occupied by the grimworm offspring of other bumblebeetles – rarely does a dead flish remain unclaimed. Those male grimworms which survive the journey from flish to flish without dehydrating or being eaten by other desert creatures are the ones that eventually mate. They do so with a fully-grown female they find in other flish. In this way, the species ensures that its gene pool remains as large and healthy as possible.

left
The desert hopper can eat the toughest plants of the Rainshadow Desert, finding nutrition in the most unpalatable-looking growths.

Because they are sexual and able to breed, grimworms can hardly be considered typical insect larvae. In the Human era, the larva usually did most of the eating and growing while only the adult, or imago, reproduced. Sometimes the imago phase was as short-lived as that of the bumblebeetle. Hatching, mating, egg-laying and dying all occurred within a few hours. The life cycle of an adult bumblebeetle is far simpler. Its sole purpose is as a transporter. In the bumblebeetle-grimworm cycle, it is the grimworm phase that breeds. Throughout evolutionary time, insects have been versatile enough to evolve lifestyles to survive the harshest conditions. The bumblebeetle is a remarkable example of such strategic evolution. It has combined several strategies and lifestyles to overcome the hostile environment in which it lives.

Once she has mated, the female grimworm continues to grow, consuming the last of the flish's desiccated flesh. She even eats the male grimworm after copulation.

When every source of food is finally exhausted, the female grimworm pupates, forming a tough, leathery cocoon that fills the old flish skin. Inside the pupa, the larva metamorphoses into a winged adult, which eventually hatches. By the time the bumblebeetle emerges from her pupa, already pregnant, there will be new flishwrecks to find. She will immediately set off on her brief but wide-ranging migration, in search of a food source for the next generation.

As the bumblebeetle shows, an essential condition for life in the Rainshadow Desert is the ability to range over large areas looking for food. This requirement has been solved in different ways by different animals, but perhaps the most surprising, when we consider its ancestry, is the desert hopper.

Night is falling. All day, the sun has heated the sands of the Rainshadow Desert. Now, the air begins to cool and the mountains cast long shadows. At the edge of the plain,

the NORTHERN FOREST

THE REGION IN THE NORTHWEST OF PANGAEA II is pounded by saturated onshore winds. The global circulation of the atmosphere brings constant westerlies to this part of the landmass, winds which travel over a vast surface of warm ocean, filling up with moisture as they go. Thick black clouds roll in from the sea and condense into water as soon as they reach land. Sunlight is rarely seen here. Rain falls relentlessly from the overhead darkness, drenching everything below.

With no mountain range to act as a windbreak, the rain-sodden region stretches hundreds of miles inland. Great rivers carry the runoff back to the sea through swamps and lakes surrounded by deep, murky forests. With all this water, a carbon dioxide-rich atmosphere and warm global temperatures, forests thrive and grow to immense proportions. The tallest trees here are conifers, which grow to the same great heights as the giant redwood trees that have dominated this area since the Triassic period.

Only a handful of specialized species are able to survive the wet conditions of the Northern Forest. Angiosperms – or flowering plants – are rare in this lush forest. They have been replaced by another highly versatile organism: lichen. Lichens are the result of a symbiotic association between algae and fungi, their cells growing in such close proximity that each gains sustenance from the other. The fungus provides a protective structure for the algae, while the algae synthesize and secrete carbohydrates as a food source for their host.

Human-era lichens were small, low-growing organisms. In the moisture-laden atmosphere of the Northern Forest, they have prospered, evolving into sturdy, tree-like forms. Gone are the soft fleshy bodies of their ancestors and in their place have developed robust trunks built up from dead fungal fibers laid down in the core. The understory of the Northern Forest is a tangle of such lichen trees. They achieve photosynthesis and absorb moisture by trailing feathery algal structures that hang like tattered curtains in the humid air. Spore sacs contain assemblages of both lichen and fungal spores, which explode on contact. The dispersal of these spores is aided by animals as they brush past the lichens, bursting the spore sacs as they go. In the rich habitat of the Northern Forest, there is no shortage of animal life.

"In the northwest corner of the supercontinent, there is a vibrant rainforest. The forest is dominated by conifers, that is, pine trees and their relatives. Flowering plants are largely extinct and have been replaced by lichens, which have evolved the capacity to grow upright into small shrubs or trees. These lichen trees fill the understory of the forest."

Professor Bruce Tiffney
Paleobotanist
University of California

left

The humid air of the Northern Forest supports many forms of life, in stark contrast to the barren expanse of the Central Desert.

About 430 million years before humankind, the first marine animals made the transition from ocean to land. The first animal colonists were invertebrates: arthropods (animals with an external skeleton and jointed limbs) and their relatives. They were able to move without the natural buoyancy of sea water using muscles within their hard outer skeletons, or exoskeletons. They had no need for lungs, taking oxygen from the air by diffusion, through pores in the cuticle. Next out of the sea were vertebrates, bony fish which had evolved primitive lungs. Over time, their fins would develop into the legs and arms of all four-legged vertebrates.

Despite their early emergence from the sea, invertebrates were not equipped to grow to any great size. They not only lacked the muscle power to support their weight, but diffusion was not an effective method of delivering oxygen into a large body. Vertebrates, with their strong internal skeletons and efficient lungs, had fewer restrictions on size. They dominated the land for 400 million years.

The last of the land vertebrates died away in the mass extinction, 100 million years after humankind. An important evolutionary niche had been vacated, and it was filled by cephalopods – octopuses, cuttlefish and squids.

To successfully colonize land, cephalopods had to improve their land-living adaptations. Their arms became stronger, the muscle fibers forming a criss-crossing network that could contract to produce a solid support or extend into flexible limbs. Over time, the arms developed beneath the body as powerful columnar legs. Since these legs could support more weight, larger animals could evolve. Meanwhile, the cavity below the domed mantle developed into a type of lung. At last, cephalopods had discovered an efficient means of breathing air.

With such adaptations, a new group of land-living squid, called terasquids, evolved. They branched into many forms, from small hunters to the biggest land animals of the time. The biggest of all is the megasquid.

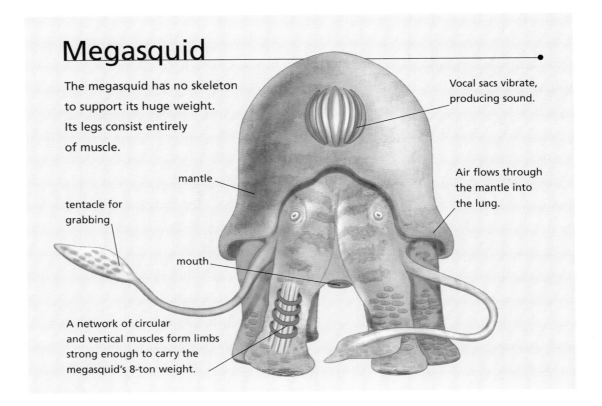

Megasquid

The megasquid has no skeleton to support its huge weight. Its legs consist entirely of muscle.

tentacle for grabbing

mantle

mouth

A network of circular and vertical muscles form limbs strong enough to carry the megasquid's 8-ton weight.

Vocal sacs vibrate, producing sound.

Air flows through the mantle into the lung.

left

The colossal megasquid, roughly the size of a Human-era elephant, crashes through the Northern Forest on powerful legs.

149

Heavier than an elephant and almost as tall, the megasquid pushes its way through the soaking vegetation, splintering conifer trunks and pulping the branches of lichen trees as it goes. Air passing through its body causes internal membranes – the megasquid's vocal chords – to vibrate and sound is produced. The sound is amplified by means of a balloon-like sac on the squid's forehead which expands and vibrates like the throat of a frog. As it moves through the forest, the megasquid stakes its territory by bellowing out signals through the drumming rain.

Eight legs, each as broad as a tree trunk, support the beast as it lumbers through the forest, moving at speeds similar to a human walking pace. When the animal is at rest, its legs are compressed and solid. An interlacing network of cartilage within each column locks together, forming a rigid support for the enormous weight of the animal. In motion, the limbs are lengthened, raising the megasquid from the ground and allowing alternate pairs of legs to swing forward. Walking on eight legs isn't easy, but the squid achieves it with a slow, fluid gait.

A pair of flexible tentacles reaches out in front of the megasquid, extending to just over six feet (roughly two meters) in length. With their dextrous tips, these limbs gather fruit, young leaves and shoots, as well as tree-dwelling creatures and their nests. Food is then delivered to the squid's mouth, at the center of its ring of legs. Despite its size, the megasquid's food requirements are small. Being cold-blooded, it does not need to burn food to maintain its body temperature. It therefore only needs about one tenth of the amount of food that a similarly sized, warm-blooded creature, such as the African elephant, would have required. Consequently, the wet Northern Forest can support large populations of megasquid.

The megasquid is not terribly clever. It doesn't have to be. With plenty of food so readily available, an absence of predators and movement achieved by the simplest of processes, the brain of a megasquid is tiny. It weighs about one pound (about 400 grams) – a tiny fraction of its eight-ton bodyweight. Not all species of terasquid are stupid, however.

The slithersucker

The slithersucker is a slime mold, a very primitive form of life which is neither animal nor plant, but a large, organized community of microbes. The slithersucker lives in the lichen trees of the Northern Forest and is an efficient predator. At certain times of day, it oozes along a branch and dangles strands of itself below, forming a sticky curtain. A passing forest flish is easily trapped in the slithersucker's slimy net. Once the flish has been caught, the slithersucker slides off the branch and crashes to the forest floor. There, it secretes a digestive acid which slowly dissolves the helpless forest flish.

The slithersucker, a shape-shifting slime mold, spreads its slimy trap for unwary forest flish.

left

A baby squibbon passes food to its sibling. With their stereoscopic eyes, big brains and prehensile tentacles, squibbons are the acrobats of the Northern Forest.

In the trees above the megasquid, a party of squibbons swings about, keeping wary eyes on the big animal below. Squibbons are the most agile of the terasquids, having adapted their natural dexterity to a tree-dwelling existence. They swing through the branches, not brachiating (swinging one arm after the other), as gibbons used to, but looping end-over-end in a continuous somersaulting action. Their eyes are set on muscular stalks, which stay with the body's center of gravity as they swing through the trees, always looking forward towards the next branch. With their sharp vision and large brains, they are able to navigate through the forest at speed.

Squibbons feed mostly on plants, but are agile enough to snatch forest flish from the air, grabbing their prey with a pair of dexterous tentacles. Each tentacle is equipped with highly-developed suckers, forming finger-like protruberances which are so flexible that a squibbon can manipulate small objects and even use simple tools.

Squibbons live in communes, building simple, nest-like structures in the uppermost branches of the forest. They are highly sociable animals, working together to defend their territory from any megasquid that blunders into it. Should one of these giants seize a baby squibbon in its suckered tentacles, it can expect to be mobbed and harassed until the prey is released. Branches are used as clubs and missiles to fend off the huge predator. Indeed, weapons and warfare seem to be evolving again after 200 million years.

Squibbon society displays an intelligence closer to that of humans than anything that has evolved since the Human era. While the ability to operate tools and act communally reflects an intelligence ideally suited to life in the Northern Forest, it may be that a changing environment will encourage the development of even greater sophistication. Perhaps a reasoning type of intelligence will evolve once again.

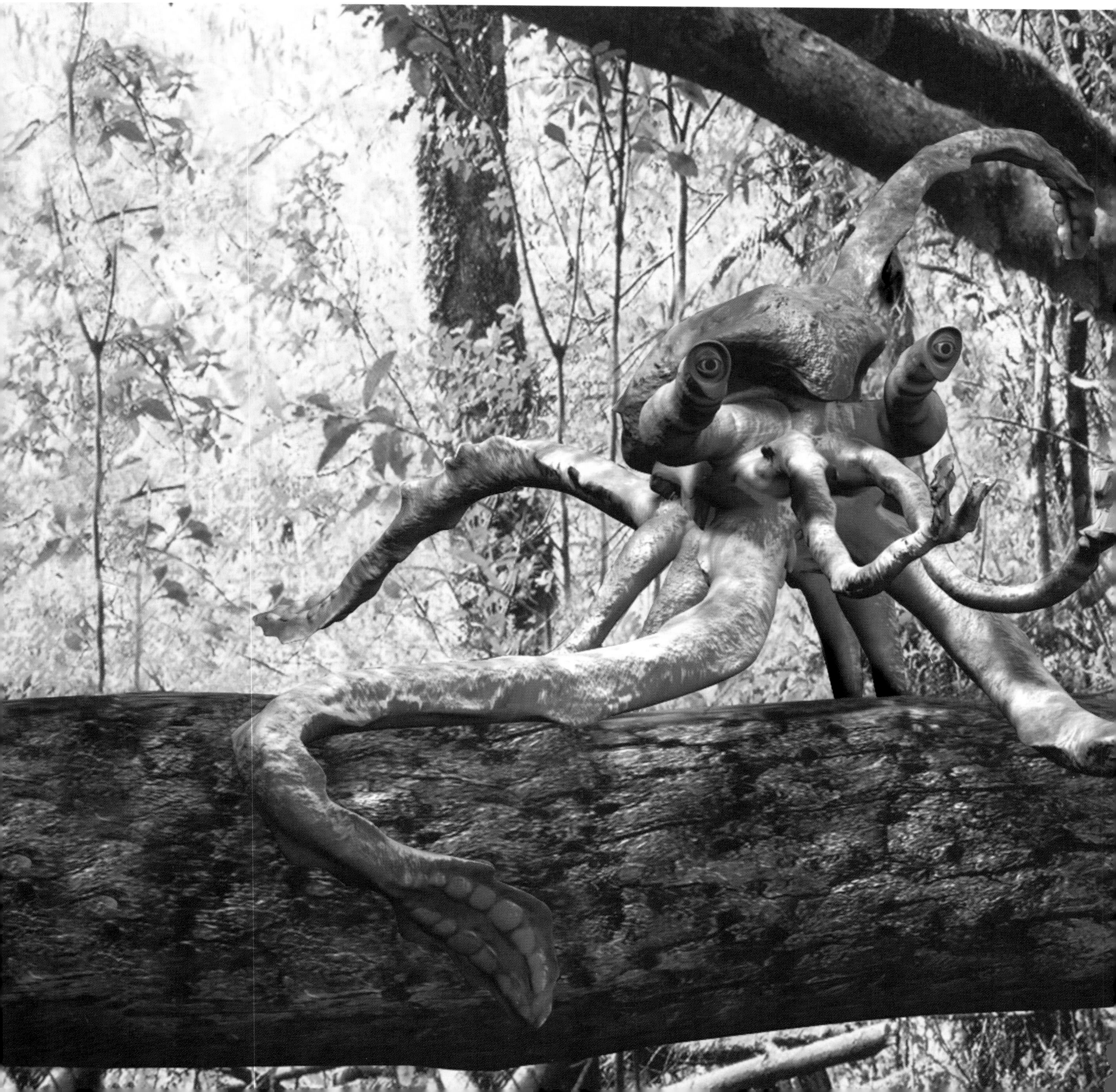

eusocial Term used to describe organisms, such as termites, that live in a colony and have highly-developed social relationships.

evolution The natural processes by which organisms change from generation to generation over time, giving rise to the diversity of life.

exoskeleton A hard outer skeleton such as that found in **arthropods**.

family *see* **taxonomy**

food chain A hierarchy of living organisms in which each eats the one below it in the chain. For example, carnivores may eat certain herbivores, which in turn eat certain types of plant.

food web A set of interconnecting **food chains** by which energy and materials circulate within an **ecosystem**.

formic acid A corrosive acid produced by certain insects. An ant bite contains formic acid.

freeze-thaw The periodic freezing and thawing of ice in rock fissures, causing erosion.

fungi (*singular* **fungus**) A family of single-celled or multicellular organisms not containing chlorophyll. Fungi reproduce by dispersing spores and live by directly absorbing nutrients from organic matter. A mushroom is a fungus.

gamete A reproductive cell.

genus *see* **taxonomy**

gizzard A type of muscular stomach found in certain birds, invertebrates and fish.

glacier A large body of ice formed from compacted snow which slowly flows out across a landmass.

gryke A vertical crack in limestone caused by weathering and water action.

gyre A large-scale system involving the circular motion of ocean currents.

hydrosphere The water component of Earth's surface, including oceans, rivers, lakes and water vapor in the atmosphere.

hypersaline Relating to water containing a high concentration of salt.

icecap An extensive dome-shaped mass of ice, spreading from a center and covering a large area. The North and South Poles are icecaps.

imago The adult form of an insect.

invertebrate An animal without a backbone, such as a centipede.

isthmus A narrow strip of land connecting two large areas of land.

karst A limestone landscape characterized by caves, gorges and subterranean streams.

keratin A hard, fibrous protein occurring in the **epidermis** of reptiles and mammals, and forming scales, claws, hair and horns.

kingdom *see also* **taxonomy** A major taxonomic division into which all known species are divided. Modern taxonomy recognizes five kingdoms: Animalia (animals), Plantae (plants), **Fungi**, Prokaryotae (**bacteria**), and Protoctista (**algae**).

larva (*plural* **larvae**) The juvenile form of an insect, often referred to as a grub.

lichen An organism made by the symbiotic association of **fungi** and **algae**.

lithosphere The rigid outer layer of Earth, comprising the **crust** and upper **mantle**.

magma Hot, molten rock found below Earth's **crust**.

mantle (1) The interior of Earth between the **crust** and the core. (2) Soft exterior membrane of the body of a **mollusk**.

mass extinction The death of many species in a short period of time.

microbe A microscopic organism, such as a **bacterium**.

mimicry The resemblance of one organism to another or to an object in its habitat for concealment or protection against predators.

Also collective, or co-operative, mimicry, in which a group of animals co-operate to mimic the appearance of another animal.

mollusk A large **phylum** of soft-bodied, **invertebrate** animals, usually with a hard shell, including snails, oysters, octopuses and squid.

mustelids A **family** of mammals including weasels, otters and wolverines.

mutation A random change in the genetic material of an organism. Mutations can lead to the formation of new **species**.

natural selection The process by which organisms pass on beneficial, or neutral, genetic characteristics to their offspring. As a result, certain genetic characteristics are eliminated, while others remain in play, accounting for the diversity of life on Earth.

neoteny The persistence of juvenile features in the adult form of an organism or the attainment of sexual maturity while still in a larval stage. For example, a certain **species** of salamander retains the gills of the larva in the adult.

niche The position or role of a **species** within its community, determined by factors such as the type of food it consumes, its predators, the climate it can survive in, etc. A lion, for example, is a meat-eater, has no predators (except for humans), and lives in open areas such as savannah. This is its niche.

nicitating membrane A thin fold of skin beneath the eyelid of reptiles, birds and certain mammals used to clean or protect the eye. Also known as a 'third eyelid'.

nudibranch A shell-less **mollusk** commonly termed a sea slug. Usually brightly colored to warn predators of its toxic properties.

order *see* **taxonomy**

outwash A mass of rocky debris deposited by a melting glacier.

Pangaea An ancient continent which formed about 300 million years ago and began to break up about 200 million years ago.

parthenogenesis A form of reproduction in which the female **gamete** develops into a new individual without fertilization by a male gamete. Also known as virgin birth.

pectoral fins A pair of fins just behind the head of a fish, used for stability and steering.

permafrost A layer of permanently frozen soil, generally found in **tundra** regions.

photosynthesis The process by which plants produce simple sugars from carbon dioxide and water, using sunlight as an energy source.

phylum *see* **taxonomy**

pingo A dome-shaped hill with an outer layer of soil covering a core of solid ice, formed in regions of **tundra** where water springs through the **permafrost**.

plankton Microscopic organisms that drift in the surface waters of the ocean. A major source of food for marine animals.

pollination The process of transferring pollen from the stamen, or male part of a flower, to the stigma, or female part of a flower. In **self-pollination**, the pollen is transferred from the stamen to the stigma of the same flower. In **cross-pollination**, pollen is transferred from one flower to another on the same plant or to a flower of another plant of the same **species**.

polychaete A segmented marine worm.

polymer A large molecule made up of a chain of simpler repeating units. Natural polymers include starch and cellulose.

polyp A marine **invertebrate** with a tubular body and a mouth surrounded by tentacles. One example is a sea anemone.

prehensile Adapted for seizing, grasping or holding. A monkey has a prehensile tail.

pupa (*plural* **pupae**) A stage in the life cycle of certain insects in which the **larva** becomes enclosed in a protective cocoon. Inside the cocoon, it undergoes major developmental changes to emerge as an **imago**, or adult.

pycnogonid A **class** of sea spiders.

pyroclastic flows A mixture of superheated rock fragments and hot gases produced by a volcano during an eruption.

quadruped An animal with four limbs specialized for walking.

radula A flexible, tongue-like organ in certain **mollusks**, having rows of teeth for feeding.

rainshadow desert A desert formed in the lee of mountains which block access of moisture-bearing winds.

rhinophores A pair of chemical sensors used by sea slugs to detect food.

savannah Large expanses of open grassland, typical of present-day tropical Africa.

self-pollination *see* **pollination**

siphonophores An **order** of jellyfish, each consisting of a colony of individuals and including the Portuguese man-of-war.

slime mold A multicellular organism that forms a small, slimy body and produces spore-bearing reproductive organs.

species *see also* **taxonomy** Members of an animal or plant species share similar characteristics and are capable of interbreeding to create fertile offspring.

stereoscopic Type of vision where two forward-facing eyes are used to see space three-dimensionally.

stratum (*plural* **strata**) A layer of rock.

subduction The process by which one **tectonic plate** slides beneath another, drawing old **lithosphere** down into Earth's **mantle**. The area between plates where this occurs is known as a subduction zone.

swim bladder A gas-filled sac used by bony fish to regulate buoyancy.

symbiosis A close association where two **species** are dependent on one another. For example, flowering plants rely on insects for pollination, while the insects feed on nectar from the flowers.

taxonomy The science of classifying living organisms. In the taxonomic hierarchy, each **species** belongs to a **genus**, each genus belongs to a **family**, and so on through **order**, **class**, **phylum** and **kingdom**, as shown below.

Common name	Dog	Housefly
Kingdom	Animalia	Animalia
Phylum	Chordata	Arthropoda
Class	Mammalia	Insecta
Order	Carnivora	Diptera
Family	Canidae	Muscidae
Genus	Canis	Musca
Species	Canis familiaris	Musca domestica

tectonic Relating to structural movements and processes in Earth's **crust**.

tectonic plates Large, rigid blocks of lithosphere that make up Earth's surface.

thorax The part of an insect's body, between the head and the abdomen, bearing the wings and legs.

trade winds Prevailing winds blowing towards the equator.

tundra A large treeless zone in the present-day northern hemisphere, characterized by a cold climate and **permafrost**.

vascular system A system of vessels or specialized tissues used to circulate liquids within an organism.

vector An organism that acts as a carrier, transferring material between other organisms. For example, insects carrying pollen between plants play a vital role in the fertilization process.

vertebrate Any animal with a backbone. Humans, horses and hummingbirds are all vertebrates.

INDEX

Page numbers in italics refer to pictures.

a

acid rain 106, 123
adaptive radiation 17, 89, 121, 154
Africa 64, 79, 111
African Plate 35
agamid 36, *36*, 154
albatross 90
algae 36, 89, 145, 154
    algal farming 74, 76, 116-118
    algal reefs *66*, 67-70
    symbiosis 67, 74, 76, *114*, 116-118, *116*, *119*
Alps, European 35
altruism 53, 59
Amazon Basin 45, 48
Amazon Grassland 24, *25*, *44*, 45-51
Amazon river 45
amphibians 9
Andes 135
angiosperms 145, 154
Antarctic Forest 64, *65*, 89-94
Antarctica 64, 65, 89, 90, 106, 111
ants 116
aphids 9
Appalachian Mountains 53
arachnids 104, 154
arthropods 92-93, 104, 124, 149, 154
Asia 64, 67, 79, 97
asthenosphere 12-14, *12*, *14*, 154
Atlantic Ocean 53
atmosphere 12, *12*, 60, 64,104, 123
    carbon dioxide 60, 97, 145
    global circulation 145
    humidity 24, 46, 53, 60, 64, 79, 81, 145
    oxygen 92, 93
    volcanic activity 14, 60, 64, 97, 106
Australia 64, 97, 111

b

babookari 46-47, *46*, *47*, 48, 50
bacteria 18, 36, 84, 86, 121, 154
    symbiosis 18, 86, 129
badger 42
bamboo 98
bat 9, 17, 57, *57,* 59, *59*
    wings 59, 127

Batesian mimicry 92
bees 69
beetle
    blister 94, *94*
    bombardier 91, *91*
    bumblebeetle *136-137*, 137-140, *138*, *139*, 142
    spitfire 93-94, *95*
    tortoise 94
Bengal, Bay of 64, 79
Bengal Swamp 64, *65*, *78*, 79-86, 106
bioluminescence 129, 132, *132*, *133*, 154
biomechanics 8
birch 29, 35
birds 9, 17, 89-92, 98, 126
    burrowing 55-59, *55*, *56*
    flightless 32, 48, 50, 55, 90
    migrant 28, 90
    of prey 41, 48, 98
    sea 32, *32*, 89-90, 126-127
    wings 17, 59, 98-99, *127*
blubber 32
boar, wild 40
body size and temperature 29, 54
bones 8, 149
brachiate 151, 154
breathing 9, 82, 146, 149
brine fly 36, *37*
brine lakes 35, 36, 40
bristleworm 118, 121
budding 70
bumblebeetle *136-137*, 137-140, *138*, *139*, 142
buoyancy 68, 70-71
burrowing species 29, 55-59, 93
bushfires 45, 48, 50-51
buzzard 59

c

calcium 67, 74
Cambrian period *20*
camouflage 29, *30*, 31, 36, 37, 80, *80*, 84, 128, 132
Canada 24
caracara 48
carakiller 48, *49*, 50, *50*, 51
carbon dioxide 60, 64, 97, 121, 145
Carboniferous period *20*, 79, 92

carnivorous plants 18, 142, *143*
castes 103, 105, 114, 116-117, 154
catfish 80
caves 41, 113, 121
Cenozoic era *21*
Central America 45, 54
Central Desert 110, *111*, 113-121
cephalopods 82-84, 128, 149, 154
cheetah 47
chemical defenses 91, *91*, *93*, *114*, 116
chromatophores 84, 128-129, 154
climate change 19, 60, 106
climatology 8
clints 40-41, 154
co-evolution 84
coastal deserts 113
cocoon 140
cod 126
cold-blooded species 80, 86, 126, 150
commensalism 18, 154
communal organisms 70-71, *70*, 75-76, 150, *150*
communication 47, 83, 128, 129, 132, 146, 150
conifer 29, 35, 45, 145, 154
continental crust 14, *15*
continental deserts 113
continental drift 8, 12-15, *13*, 64, 65, 79, 89, 97
continental plates 12-15, 35, 64, 65, *65*, 79, 89, 97, 111, *111*
continental shelves 27, 60, 135, 154
convection currents *12*, 14, 154
coral 67
core, Earth's 12, *12*
cotton grass 27
crab 9, 124
crane 98
Cretaceous period *21*
Crete 35
crocodile 81
crust, Earth's 12-15, *12*, *15*, 64, 97, 154
crustaceans 104, 154
cryptile lizard 36-37, *37*, *38-39*, 40
cuticle 91, 92, 114, 142, 149
cuttlefish 83, 128, 149
Cyprus 35

Picture credits

Ardea: Thomas Dressler, 48 • Corbis: David Muench, 52 • Thomas Eisner, 91 • Kevin Flay, 46, 47, 49, 51 • Getty Images: FPG/Richard H Johnston, 122 • VCL, 19 • Image Bank/Richard A Brookes, 19 • Michael Melford, 134 • Stone/David C Tomlinson, 18 • John Hafernik, 94 • Los Angeles County Museum of Natural History, 132 • Tom Mackie, 34 • Naturepl.com: Ingo Arnat, 106 • Peter Blackwell, 16 • Juan Manuel Borrero, 88 • Fabio Liverani, 82 • Steven D Miller, 78 • Peter Oxford, 17 • NHPA: Bill Coster, 32 • Stephen Dalton, 59 • Daniel Heuclin, 112 • B Jones & M Shimlock, 69 • Rich Kirchner, 52 • T Kitchin & V Hurst, 29 • Norbert Wu, 18 • Steve Nicholls, 8-9, 20-2, 26, 33, 35, 44, 96, 144; backgrounds 11, 28, 43, 46, 47, 49, 51, 72-3, 77, 84-5, 87, 93, 95, 99, 100-1, 115, 140, 148, 150, 151 • OSF Films: Kathie Atkinson, 70 • Waina Cheng, 141 • Judd Cooney, 42 • Rudi Kuiter, 128 • Zig Leszczynski, 81 • Richard Manuel, 118 • Colin Monteath, 14 • Scott Winer, 108-111 • Belinda Wright, 36 • Steve Weston, 83.